Leonardo Da Vinci's Water Theory

On the origin and fate of water

by

L. Pfister
Department of Environment and Agro-Biotechnologies
Centre de Recherche Public – Gabriel Lippmann
Grand-Duchy of Luxembourg

H. H. G. Savenije
Department of Water Management
Delft University of Technology, The Netherlands

F. Fenicia
Department of Water Management
Delft University of Technology, The Netherlands

IAHS Special Publication 9

 The National Research Fund of
Luxembourg provided financial support
for this publication

This publication was financially supported by UNESCO's Hydrological
Processes and Climate Section, Division of Water Sciences and is a
contribution to the International Hydrological Programme.

Published by the International Association of Hydrological Sciences 2009

IAHS Special Publication 9
ISBN 978-1-901502-34-3

British Library Cataloguing-in-Publication Data.
A catalogue record for this book is available from the British Library.

© **IAHS Press, 2009**. All rights reserved.
Neither this book, nor any part of it, may be reproduced, stored in a retrieval system or transmitted in any form or by any means, electronic, mechanical, photocopying, recording or otherwise, without specific written permission from the publisher. No use of this publication may be made for electronic publishing, resale or other commercial purposes without the prior written permission of IAHS Press.

IAHS is indebted to the employers of the Authors for the support and services provided that enabled them to carry out their task.

The information, data and formulae provided in this volume are reproduced by IAHS Press in good faith and as finally checked by the author(s); IAHS Press does not guarantee their accuracy, completeness, or fitness for a given purpose. The reader is responsible for taking appropriate professional advice on any hydrological project and IAHS Press does not accept responsibility for the reader's use of the content of this volume. To the fullest extent permitted by the applicable law, IAHS Press shall not be liable for any damages arising out of the use of, or inability to use, the content.

The designations employed and the presentation of material throughout the publication do not imply the expression of any opinion whatsoever on the part of IAHS concerning the legal status of any country, territory, city or area or of its authorities, or concerning the delimitation of its frontiers or boundaries.

The use of trade, firm, or corporate names in the publication is for the information and convenience of the reader. Such use does not constitute an official endorsement or approval by IAHS of any product or service to the exclusion of others that may be suitable.

IAHS Publications are available from:
IAHS Press, Centre for Ecology and Hydrology, Wallingford, Oxfordshire OX10 8BB, UK
tel: +44 1491 692442; fax: +44 1491 692448; e-mail: jilly@iahs.demon.co.uk

Printed in Eynsham, England, by Information Press.

To Elodie

For proving to me every day that beyond darkness there is light, hope and joy.

Laurent Pfister

February 2009

Contents

Foreword	by Jeffrey J. McDonnell	vii
Preface	by Laurent Pfister & Hubert H. G. Savenije	ix

1	**INTRODUCTION**	**1**
1.1	Leonardo Da Vinci	3
1.2	Hydrology before and after Leonardo	6

2	**LEONARDO'S 'TREATISE ON WATER'**	**11**
2.1	Leonardo and the "*science of water*"	13
2.2	Leonardo's unfinished "*Treatise on Water*"	15
2.3	The conceptual foundations of Leonardo's "*Treatise on Water*"	18

3	**THE ATMOSPHERE**	**25**
3.1	Structure, constituents and processes	27
3.2	Meteorological phenomena	28
3.3	Measuring devices	31

4	**THE PHYSICAL STRUCTURE OF THE EARTH**	**35**
4.1	The earth as a support for the water cycle	37
4.2	The earth as a living body	37
4.3	The structure of the earth	39
4.4	Water as a force eroding and shaping the earth's surface	41
4.5	The role of sedimentation in the building of landforms	45
4.6	Leonardo's concept of erosion and the forces shaping the earth's surface, as opposed to the biblical deluge theory	47
4.7	Erosion as a constant remodelling force of the earth's surface under the effect of tectonics	49

5	**THE WATER CYCLE**	**51**
5.1	The general framework for Leonardo's concept of the water cycle	53

5.2	Leonardo's key questions related to the water cycle		54
5.3	Leonardo's concept of the water cycle		55
	5.3.1	The idea of drainage basins	55
	5.3.2	The hydrological cycle	56
	5.3.3	Leonardo's thoughts on 'where does the water go when it rains?' and 'what flow path does it take to the stream?'	59
	5.3.4	The water cycle: salinity of the sea water	63
	5.3.5	Leonardo: the first experimental hydrologist?	63
	5.3.6	On the origin of tides and the link with Leonardo's theory of the hydrological cycle	68

6   THE STUDY OF WATER IN MOTION   71

6.1	Leonardo's motivation for studying moving waters	73
6.2	Training rivers, or how to raise the economic value of rivers	75

7   LEONARDO'S LEGACY   85

References   89

Postscript   by Laurent Pfister & Lucien Hoffmann   91

List of Figures   93

Foreword

We know more about celestial bodies than soils underfoot.
Water is the driving force of all nature.   Leonardo Da Vinci (1452–1519)

The International Association of Hydrological Sciences (IAHS) is the oldest international hydrological society (now almost 90 years old), with national representatives in 98 countries, and more than 5000 individual members. While the Association has been active in many areas of the science for many decades (through conferences and symposia, coordination with the UNESCO International Hydrological Programme, etc.), the role of IAHS as the key international water organization has grown enormously as the role of water in earth system science has grown in recognition. IAHS currently leads the Decade on Prediction in Ungauged Basins, an effort to fundamentally change the field of hydrological science from calibration-based modelling to new approaches focused on fundamental understanding of hydrological systems. IAHS is also expanding its publishing model, from strictly one of Proceedings and Reports (Red Books) publishing to one that includes new book series, like the Benchmark Papers in Hydrology. It is in this context of new publishing models that the IAHS volume on *Leonardo Da Vinci's Water Theory* by Pfister, Savenije and Fenicia, is introduced—a volume brimming with new insights into the basic origins of hydrology and a timely counterpoint for the many new initiatives and discoveries within IAHS.

Pfister, Savenije and Fenicia present a dazzling array of insights into hydrological processes by hydrology's first, and perhaps most, significant figure. It is amazing to consider that as a scientist 200 years before Newton and 300 years before Linnaeus, Leonardo Da Vinci was doing hypothesis-driven science and describing and classifying myriad hydrological processes and basic properties of water. As perhaps the world's first experimental hydrologist (among his many other possible titles), Da Vinci was working on the hydrological cycle 200 years before Pierre Perrault's field studies of the hydrological cycle and Edmund Halley's experiments on evaporation. Pfister, Savenije and Fenicia note that Da Vinci came very close to the modern definition of the hydrological cycle when he explained that water passes through the major river systems countless times, summing up to volumes much greater than those contained in the world's oceans. Da Vinci's work, and in particular his contributions of a hydrological nature—summarized here for the first time—forms the hydrological bedrock of our science.

The team of Pfister, Savenije and Fenicia is a perfect trio for the Da Vinci hydrological historical narrative. Laurent Pfister is a research hydrologist at the Public Research Center–Gabriel Lippmann Institute in Luxembourg and a long-time Da Vinci-phile and avid water history buff. Huub Savenije is Professor of Hydrology at TU Delft and recent recipient of the European Geophysical Union Darcy Medal, based on his considerable contributions to international hydrology. Fabrizio Fenicia is a native Roman and a hydrological modeller at TU Delft with a gift for translation and interpretation of the many Da Vinci Codices and other obscure materials. Together, this trio of hydrological sleuths has uncovered a vast array of hitherto unassembled material and provided an impressive foundational document for hydrology. Fascinating, inspiring, revealing and amazing, *Leonardo Da Vinci's Water Theory* opens up a new history to the study of water. Pfister, Savenije and Fenicia's obvious passion for the material, the interesting modern hydrological backdrop for the materials, and accurate and careful reporting of the facts—many unearthed for the first time in this volume—make this a benchmark work.

Jeffrey J. McDonnell
Richardson Chair in Watershed Science
Oregon State University, USA

Preface

Leonardo Da Vinci (1452–1519) was not only one of the greatest artists of his time, he was also a great engineer and scientist. Only relatively recently, at the beginning of the 19th century, did we recover fragments of his many scientific writings. A large part of his scientific work was dedicated to understanding the movement, circulation and physical characteristics of water in its different forms: as a gas (water vapour), as mist, as drops, as stagnant or flowing water, as ice, hail and snow. This book aims to make Leonardo Da Vinci's contributions to the science of water accessible to a wider public and to compare his ideas with our present state of knowledge.

With the benefit of hindsight, we now know that many of the things that Leonardo Da Vinci wrote were incorrect, while in some cases he built on the false premises of classical authors who preceded him. But what makes Leonardo Da Vinci unique is his scientific approach. He can be regarded as the first hydrologist who formulated hypotheses on the basis of empirical evidence, which he subsequently tried to falsify or test under different conditions (Pfister & Savenije, 2006). Being an artist, a philosopher, an engineer and a scientist, he was capable of combining his talents for observation, for capturing images in drawings, for designing instruments to test his hypotheses, and for translating these into causal relationships. On top of that, he tried to translate his theories into quantitative relationships, difficult though this may have been in his time, given the limited means available for hydrological observation. He was probably the first hydrological experimentalist to design and build his own instruments to test his hypotheses.

Although his writings were never formally published, his work, like that of the classical Greek philosophers before him, should be considered as a benchmark in the science of hydrology. With this book, we hope that this knowledge will be brought to the community of hydrological scientists, and to all who have an interest in this great artist, scientist and engineer.

Many people have written about Leonardo Da Vinci, and recently, probably as a reaction to the novel by Dan Brown (*The Da Vinci Code*), he has received a lot of attention from the public and the media. However, little has been written about Leonardo's pioneering work on water. Hydrology being a relatively young science, the hydrological community has not spent much time looking back at how the ancient philosophers and scientists regarded water, or putting hydrology into an historical perspective.

This book aims to fill that gap, and to give Leonardo Da Vinci the credit for being one of the first hydrologists, if not the first, who used experiment and deduction as powerful tools for the development of theory. He was a real *Homo universalis* who not make a distinction between the different disciplines, science, engineering, philosophy and art. He considered *"painting as a science"*, and maybe that was his great strength, both as an artist and a scientist.

The citations taken from Leonardo Da Vinci's notebooks and used throughout this volume mainly stem from earlier publications in English by Jean-Paul Richter (1888) and by Edward McCurdy (1942, re-edited in French in 1987 and 1989). Jean-Paul Richter's compilation of Leonardo's manuscripts was originally translated from Italian into English by R. C. Bell and E. J. Poynter, while Edward McCurdy's work was translated from Italian and English into French by Louise Servicen for the re-editions of 1987 and 1989. Given the fact that these translators may well have included their own successive interpretations into Leonardo's writings, the authors have preferred to rely to a certain extent on their own verifications by referring to the original Italian text (Majer, 2006).

Creating this book has been a truly international collaborative effort between researchers of the Public Research Center – Gabriel Lippmann and the Delft University of Technology, with support coming from UNESCO, IAHS and the National Research Fund of Luxembourg. The authors are most grateful to these organisations for their support and they express their thanks to Penny Perrins and Cate Gardner for their continuous assistance and truly dedicated work throughout the production process of this book.

The authors would like to conclude with a personal note. The idea for writing this book was born in late summer 2004 in the mind of Christelle Poirier during a visit to a Da Vinci exhibition at the Foundation Pierre Giannada in Martigny, Switzerland. She suggested that we should jointly write a book in which we would compare Leonardo's ideas with modern insights in the hydrological sciences. At that time she was fighting a tough battle against a terrible disease. She lost that battle only a couple of months later. As a tribute to her, we have done our utmost to stay close to her ideas.

Laurent Pfister
Head of the research unit 'Geohydrosystems and Landuse Management'
Department Environment & Agro-biotechnologies,
Centre de Recherche Public – Gabriel Lippmann, Belvaux,
Grand-Duchy of Luxembourg

Hubert H. G. Savenije
Chair of Hydrology at the Water Resources Section
Department of Water Management
Delft University of Technology, The Netherlands

1 INTRODUCTION

The Arno River valley (5 August 1473).

Gabinetto disegni e stampe della Galleria degli Uffizi, Firenze (Inv. 436E).
© Alinari Archives/CORBIS.

1.1 LEONARDO DA VINCI

In 1568, the artist and writer Giorgio Vasari wrote, in what can be considered the first biography of Leonardo Da Vinci, that *"he attracted to himself the hearts of men. And although he possessed, one might say, nothing, and worked little, he always kept servants and horses, in which latter he took much delight, and particularly in other animals, which he managed with the greatest love and patience; and this he showed when often passing by the places where birds were sold, for, taking them with his own hand out of their cages, and having paid to those who sold them the price that was asked, he let them fly away into the air, restoring to them their lost liberty. For which reason nature was pleased so to favour him that, wherever he turned his thought, brain and mind, he displayed such divine power in his works, that, in giving them their perfection, no one was ever his peer in readiness, vivacity, excellence, beauty and grace"*. Some five centuries later, White (2000) gives a wonderful key to the understanding of the *Homo universalis* Leonardo Da Vinci, when he writes that *"at the same time as he was using his skills as an artist to represent his scientific findings, he was applying his scientific learning to improve his art"*. Indeed, it appears that Leonardo used his knowledge in anatomy to improve his representations of human and animal figures, and his studies on optics helped him perfect his use of shadow, contrast and perspective, ultimately improving the realism and accuracy of his landscapes through his knowledge of geography and geology.

Before reaching excellence in so many diverse and complementary fields of art and history, Leonardo had endured set-backs in his life. His grandfather Antonio recorded in his notebook that *"there was born to me a grandson, the child of Ser Piero my son on 15 April 1452, a Saturday, at the third hour of the night. He bears the name of Leonardo"* (Dickens, 2005). Ser Piero, Leonardo's father, is known for having been a very ambitious and successful notary in Tuscany during a period when the cities of Florence and Siena were exceedingly influential in business and at the very epicentre of the Renaissance (White, 2000). From a brief liaison with a Florentine girl named Caterina, Ser Piero was given his first son, the illegitimate Leonardo. Later in his life, this illegitimacy would prevent Leonardo from attending university; nor would he be allowed to join any of the respected professions. While these circumstances might well have paved the way for Leonardo's outstanding career as an artist and as a scientist, at a later stage of his life, he looked back with some regrets: ***"If indeed I have no power to quote from authors as they have, it is a far bigger and more worthy thing to read by the light of experience which is the instructress of their masters. They strut about puffed up and pompous, decked out and adorned not with their own labours but with those of others and they will not even allow me my own"*** (Codex Atlanticus).[*]

Giorgio Vasari describes Leonardo as a *"variable and unstable"* child, saying that Leonardo *"set himself to learn many things, and when he had begun them gave them up. In arithmetic, during the few months that he applied himself to it, he made such progress that he often perplexed his master by the doubts and difficulties that he propounded. He gave some time to the study of music, and learnt to play on the lute, improvising songs most divinely. But though he applied himself to such various*

---

[*] Bold italic font is used throughout this volume to indicate text translated from Leonardo Da Vinci's notebooks. The English translations are derived from Richter (1888a,b) and McCurdy (1942a,b), or translated directly from Majer (2006).

subjects, he never laid aside drawing and modelling in relief, to which his fancy inclined him more than to anything else".

Here, Leonardo's father, Ser Piero, played a decisive role. Being well aware of his son's outstanding capacities as well as of the difficulties that his illegitimacy might cause, he attempted to find a way to offer him a decent education. Ser Piero's efforts are described by Giorgio Vasari (1550), writing that "*He* (Ser Piero) *took some of his* (Leonardo's) *drawings one day and carried them to Andrea del Verrocchio, with whom he was in close friendship, and prayed him to say whether he thought, if Leonardo devoted himself to drawing, he would succeed. Andrea was astounded at the great beginning Leonardo had made, and urged Ser Piero to make him apply himself to it. So he arranged with Leonardo that he was to go to Andrea's workshop, which Leonardo did very willingly, and set himself to practice every art in which design has a part. For he had such a marvellous mind that, besides being a good geometrician, he worked at sculpting (making while a boy some laughing women's heads, and some heads of children which seemed to have come from a master's hand), and he also made many designs for architecture; and he was the first, while he was still quite young, to discuss the question of making a channel for the River Arno from Pisa to Florence*".

Thus, in 1467, at the age of 15, Leonardo was offered an apprenticeship by Andrea del Verrocchio, an influential Florentine sculptor and painter who worked at the court of Lorenzo de Medici in Florence (Dickens, 2005). From this moment on, Leonardo travelled relentlessly, turning his attention to numerous centres of interest. From 1482 he worked as a military engineer for Ludovica Sforza, the Duke of Milan and started taking commissions for paintings. In 1495, work began on *The Last Supper*, Leonardo's second best-known work. While staying in Milan, he also started writing his notebooks, presumably in an attempt to make himself indispensable to the Duke. Since he was often short of paper, Leonardo wrote his ideas, concepts and opinions on astronomy, geography (including hydrology), geology and anatomy on literally thousands of sheets. With French troops forcing Sforza to flee around 1499, Leonardo also had to move again and was searching for a new patron.

In 1502, Leonardo started working for Cesare Borgio, the self-appointed Duke of Romagna, for whom he served as an advisor on engineering projects (Dickens, 2005). He considered Borgio as a power-crazed and war-mongering man. At that time, Leonardo began working on his most famous work, *La Joconda* (*The Mona Lisa*), probably a simple commission from a Florentine merchant called Francesco del Giocondo.

In 1505, Leonardo returned to Milan, where he succeeded in working for the French governor of Milan, Charles d'Amboise. Having been eager to please throughout his life, he finally seemed to have found a respectful patron. In 1507, he appointed a personal assistant called Francesco Melzi.

After the death of Charles d'Amboise, Leonardo left Milan for Rome in 1513, undertaking commissions for Giuliano de Medici (Dickens, 2005). In 1516, he started working for the King of France, François I.

In the last years of his life, Leonardo was well aware of the fact that most of his notes were badly, if at all, organised. Today, only his treatise on painting, entitled *Trattato della pittura*, can be considered as a completed work. The remaining patchwork of notes and drawings is a collection of notebooks that cover more than 20 years of creative work.

When he died on 23 April 1519, Leonardo left a legacy of nearly 13 000 pages of notes. His companion, Francesco Melzi, had tried to organise Leonardo's work during his lifetime, but soon realised that this task was well beyond his capabilities. Unfortunately, after Melzi's death, the documents were soon dispersed throughout Europe, since none of Leonardo's family members showed any interest in his work. It can only be guessed that, since the writing on almost all the documents left by Leonardo is back to front (so-called mirror writing), this may well have led to the judgement of his own family that these were only useless notes and of little interest to anyone. The reason why Leonardo wrote in this way is still today a matter of mostly wild speculation.

In the end, the notebooks were spread by treasure-hunters, or sold to wealthy families. Since then, many of Leonardo's writings and drawings have been either lost or destroyed, while others made their way to libraries, private collections, and even into the collection of the British royal family. Of the 13 000 pages of documents mentioned by Melzi in 1519, the existence of only 7000 is known today.

Most of Leonardo's work is now scattered throughout Europe (and the USA) and can be indexed as follows:

Codex Arundel	1478–1518	British Library, London, UK
Anatomical studies	1478–1518	Royal Library, Windsor Castle, UK
Trattato della Pittura	1480–1516	Biblioteca Vaticana, Rome, Italy
Codex Atlanticus	1480–1518	Biblioteca Ambrosiana, Milan, Italy
Codex Paris (A-M)	1484–1515	Institut de France & Bibliothèque Nationale, Paris, France
Codex Trivulzio	1487	Castello Sforzesco, Milan, Italy
Codex Forster	1487–1490	Victoria and Albert Museum, London, UK
Codex Madrid	1490–1505	Biblioteca Nacional, Madrid, Spain
Codex Turin	1505	Biblioteca Reale, Turin, Italy
Codex Leicester	1506–1510	Private ownership (Bill Gates), USA

During Leonardo's lifetime, peer-reviewed scientific publishing did not exist. Publications on experiments, findings and concepts were known only to a handful of noblemen. Today, Leonardo is mainly known for his most famous paintings, such as *The Mona Lisa* and *The Last Supper*. His studies on the anatomy of animals and Man, as well as his "flying machines", are also known to many people. However, few people know about his writings on the water cycle. His interest in the movement of water may well have arisen because, during his childhood, Leonardo witnessed some very frightening natural disasters. In 1456 the Val d'Arno was devastated by a severe storm, while 10 years later the River Arno caused massive flooding across the entire region. These events appear to have had a lasting impact on Leonardo, as we can see through his countless drawings of deluges. While fascinated by the destructive power of water, he was at the same time eager to find ways as to how water could be trained and controlled.

1.2 HYDROLOGY BEFORE AND AFTER LEONARDO

Whether it was because of the threatening and devastating power of floods or due to the need to find a reliable supply of drinking water, we know that mankind has tried for thousands of years to understand the origins of rivers and springs, as well as the causes for exceptional meteorological and hydrological events. Numerous philosophers, scientists and engineers have dedicated their attention to the science of hydrology over the past 3000 years. It is difficult to estimate how many of the writings, concepts and ideas that preceded Leonardo, were really known by him. In the following we provide an overview of the sources that were available to Leonardo. In doing so, we have made ample use of the books by Biswas (1970, 1972) *History of Hydrology*, and Brutsaert (2005) *Hydrology: An Introduction*, and the *Encyclopedia of Hydrology and Water Resources* by Herschy & Fairbridge (1998).

Around 800 BC, the Greek poet Hesiod referred to certain features of the atmospheric water cycle, such as evaporation and precipitation: "*For the morning is cold when Boreas* (the god of the north wind) *bears down; in the morning from the starry sky over the earth a fertilizing mist spreads over the cultivations of the fortunate; this* (mist), *drawn from ever flowing rivers, and lifted high above the earth by a storm wind* (reference to evaporation), *sometimes falls as rain toward evening* (reference to precipitation), *or sometimes blows as wind, while Thracian Boreas chases the heavy clouds*".

The *Chandogya Upanisad* (VI, 10), a major text in Hinduism, briefly refers in a very beautiful way to what we might today qualify as the neverending character of the water cycle: "*These rivers, my son, flow, the eastern toward the east, the western toward the west. They go from sea to sea. They become the sea itself, and while there, they do not know which river they are*". This passage even seems to refer to one of the major research topics of modern hydrology: the origin and fate of water.

Due to the importance of the Nile's floods for maintaining the fertility of the land in the Nile valley, recordings of the floods commenced at least 1000 years prior to Thales of Miletus (624–584 BC), mainly to implement and run an agricultural tax system. As far as we know the first philosophical theories about the origin of the world were developed in Greece around the 6th century BC by Thales. He considered water as being "*the original substance, and hence being the material cause of all things*". Moreover, he was convinced that "*the earth floats on the water*".

His disciple, Anaximander of Miletus (610–545 BC) considered that "*in the beginning, human, as well as animal life, originated in water*". He described rainfall as being produced by vapour that rises from the earth. He believed everything was cyclic and that we go through perpetual cycles of creation and decay, which made the world predictable (as opposed to being ruled by gods). As part of his concept, Anaximander thought that originally the earth was entirely covered by the sea. Through evaporation, this sea would ultimately disappear and in the end only dry land would remain.

Anaximenes of Miletus (585–525 BC), a follower of Anaximander, believed that clouds were generated by air that condenses, and that this condensation gave rise to rainfall, hail or snow depending on temperature conditions. In contrast to Thales who considered water as the origin of everything, Anaximenes attributed this phenomenon to air: "*Just as our soul, being air, holds us together, so do breath and air encompass the whole world*".

Xenophanes of Colophon, who lived between 570 and 470 BC, believed that *"the sea is the source of water, and the source of wind"*. He correctly explained the presence of marine fossils in high mountainous locations as being due to the fact that *"the land must have been under the sea at one time"*.

The first philosopher to have a correct perception of the hydrological cycle was Anaxagoras (500–428 BC). He thought that rivers were generated from rainfall and underground water that is stored during winter in the cavities of the earth. Also, he believed that the sea is saline because water washes the earth, extracting salts and depositing them in the sea. An interesting detail is that this great philosopher-hydrologist was banished from Athens and accused of atheism because he had stated that the stars and the planets were not divine, but made of stone.

Herodotus of Halicarnassus (484–425 BC) described Egypt as a gift of the River Nile. He understood that the alluvial land in the delta had been built up by the deposition of silt, previously eroded upstream, during the annual floods of the Nile.

Meanwhile, Democritus of Abdera (460–357 BC), like Anaximander of Miletus, had recognised evaporation and condensation as major processes in the formation of rain, and he had understood the role that these processes play in the transport of water over large distances.

The philosopher Plato (428–348 BC) had a different way to explain natural phenomena. His approach was metaphysical and religious as opposed to the scientific approach of Anaxagoras. Plato followed the concept of *Anima mundi*, and stated that the world is a living body, just like a human being. He thought that the water that generates springs, rivers, lakes and seas originates from a big cave located deep in the earth. This cave, named *Tartarus*, periodically breathes, pumping the water around the earth. Plato described water as *"flowing through subterranean channels and finding its way to several places, forming seas, and lakes, and rivers and spring"*.

Leonardo shared the ideas of Plato and believed in the correspondence between the human microcosm and the macrocosms of nature. He described nature as characterized by a vegetative soul and regarded the elements of nature, such as stone, water and conduits as related to the bones, blood, and veins of the human body. Plato believed in the theory of polyhedra, stating that there could only be five regular solid figures having as their sides, regular identical polygons without any re-entrant angles. These so-called Platonic bodies were assigned by Plato to the elements of air (octahedra), water (isosahedra), fire (tetrahedra) and earth (cube). While those elements were supposed to be constituted by these bodies, a fifth body, the dodecahedron, was according to Plato used by god *"in the delineation of the universe"*.

Leonardo gave a detailed justification for his belief in Plato's concept of the elements: ***"Of the figures of the elements; and first as against those who deny the opinions of Plato, and who say that if the elements include one another in the forms attributed to them by Plato they would cause a vacuum one within the other, I say it is not true, and I here prove it, but first I desire to propound some conclusions. It is not necessary that the elements, which include each other, should be of corresponding magnitude in all the parts, of that which includes and of that which is included. We see that the sphere of the waters varies conspicuously in mass from the surface to the bottom, and that, far from investing the earth when that was in the form of a cube that is of eight angles as Plato will have it, that it invests the earth which has innumerable angles of rock covered by the water and various*** *prominences and concavities, and yet no vacuum is generated between the earth and water; again, the air invests the sphere of waters together with the mountains and valleys, which rise above that sphere, and no vacuum remains between the earth and the air, so that any one who says a vacuum is generated, speaks foolishly. But to Plato I would reply that the surface of the figures which according to him the elements would have, could not exist*". (Codex Paris, F).

In the third century BC, Aristotle recognized the interaction of two different processes in the formation of flow. The first is the one expressed by Anaxagoras, to whom he makes reference, i.e. that rivers are the result of rainfall plus underground water. However, he reckoned that this process could not be sufficient to explain the amount of water that flows in the rivers during periods of dry weather, stating that the storage of water required would exceed the mass of the earth. Hence, Aristotle introduced a second integrative process. He noticed that the air mass condenses during the night, producing a downward flux of water, and the humidity on the earth evaporates during the day, generating an upward flux of water. By analogy, he thought that just as cold conditions generate water from the air, so they can generate water from the earth. Aristotle considered the hydrological cycle as being endless. In contrast to Plato, he did not believe in a subterranean reservoir generating rivers. As we know today, during his lifetime Leonardo's concept of the water cycle always remained close to Plato's idea of a subterranean reservoir feeding surface waters. Leonardo shared the ideas of Aristotle who believed that the world is constituted of four elements: earth, water, air and fire, which are displaced in concentric spheres with the heavier (earth) at the centre and the others around it. He also thought that the elements tend to stay with themselves, in the sense that if they are separated, they tend to come back together through gravity.

Ecclesiastes (250 BC) came much closer to a correct concept of the water cycle, when he stated that "*all the rivers run into the sea; yet the sea is not full; unto the place from whence the rivers came, thither they return again*". Unfortunately, he did not fully understand the role of evaporation and condensation in the never-ending hydrological cycle.

The Roman school of thought was greatly influenced by the Greek tradition. Lucretius Carus Titus (96–55 BC), in *De Rerum Natura*, speaks about the subject more in poetic than scientific terms. However, he combines the ideas of Aristotle regarding the evaporation and condensation of water, and of Plato on the generation of underground water and river flow. Consequently, like many of his predecessors, he did not recognize the role of rainfall in generating discharge in rivers, but believed in the filtration of seawater as the origin of rivers.

Leonardo presumably consulted the *Metamorphoses* of Ovidius (Ovid) (43 BC–AD 17), from whom he took some ideas about the biblical deluge and the succession of geological eras.

Seneca (4 BC–AD 65) also referred to the theory of Aristotle and stated that water can be generated by the soil. However, in order to explain the reason why the level of the sea does not progressively increase, Seneca explained that just as soil can be transformed into water, all the main elements, air, water, earth and fire, can be transformed into each other. Also, he considered that just as in our bodies there are veins that carry blood and air, also in the body of the earth there are conduits that transport water. Moreover, he believed that, just as blood is generated from a broken vein, water springs out of these conduits when they are broken. Here again, we find concepts that Leonardo used for his conceptualisation of the water cycle.

Leonardo read the Encyclopaedia of Pliny the Elder (23–79 AD), the *Naturalis Historia*, composed of 37 books dealing with geography, anthropology, zoology, botany, medicine, mineralogy and art history. As shown later in this book, in some of his writings Leonardo argued with Pliny, whom he used as an imaginary opponent while presenting his theories.

Marcus Vitruvius Pollio, a famous architect in the Roman era who lived in the first century BC, wrote a large 10-volume work entitled *De Architectura*, in which he demonstrated a clear view of the water cycle, resembling that of today. He speaks about rainfall, evaporation, infiltration and snow melt as the principal components of the water cycle. In his view, only the thinnest, i.e. the lightest, water could be evaporated and the heaviest remained in place. He also tried to understand groundwater dynamics, which is necessary to locate the underground conduits and springs.

In his book *De Aquis Urbis Romae, Libre II*, Sextus Julius Frontinus (AD 35–104), a commissioner of water works in Rome, provided a huge amount of information related to Roman water supply systems. Frontinus did not realise that velocity needs to be measured in order to calculate discharge in open channels. In fact it is believed that for many centuries, the design and implementation of the Roman aqueducts relied on a process of trial and error by the engineers.

However, Hero of Alexandria (first century AD) did calculate streamflow by multiplying the velocity with the cross-sectional area, and was likely the first to do so. Interestingly, this concept was not taken into consideration by Frontinus.

After the collapse of the Roman Empire until the time of Leonardo (i.e. 476 to 1450 AD), little was done on the theme of water, and most of the heritage of the past was preserved in monasteries where texts were copied or reformulated. Generally, during that period scientific studies experienced a period of stagnation in favour of studies related to religion and alchemy.

However, Leonardo was influenced by some medieval authors, such as Ristoro d'Arezzo, who in his manuscript *Della Composizione del Mondo*, dating from 1282, compared the water cycle to the circulation of blood in the human body. Leonardo referred to this concept on many occasions and even relied on d'Arezzo's view that water was lifted through a siphon effect from the lowest parts of the sea to the top of the mountains, where it would in the end generate springs and give birth to rivers.

In addition to the works by Aristotle, Plato, Pliny the Elder, Ovidius and Vitruvius Pollio, which constitute his fundamental sources, Leonardo had access to works of his time, but they were probably of less importance for the formation of his ideas. These include the *Speculum Maius* of Vincent de Beauvais (*c.* 1400 AD), which is an encyclopaedia that was widely used in the Middle Ages, the *Tractato delle più maravigliose cosse e più notabili che si trovano in le parti del mondo*, by Jehan de Mandeville (*c.* 1500 AD), and finally the *Liber secretorum, de virtutibus herbarum, et lapidum et animalium quorundam, et de mirabilibus mundi*, by Albertus Magnus (*c.* 1400 AD).

The poor progress made during the 1500 years prior to the Renaissance is illustrated by the fact that while writing a treatise on the flood of the River Tiber in Rome in December 1598, Giovanni Fontana Da Meli (1540–1614) did not use the flow velocity to determine discharge.

However, during the 16th century, hydrological concepts began to evolve with Jacques Besson (1500–1580), a contemporary of Leonardo, giving a clear and mostly correct description of the hydrological cycle. Evaporation of water by the sun, condensation and precipitation were supposed by him to be sufficient to sustain river flow and springs. At the same time, Bernard Palissy (1499–1590) was convinced that rivers and springs could have no other origin than rainfall. It is very probable that Besson and Palissy knew of each other's writings. It is very unlikely that Leonardo came across Besson's or Palissy's work.

Like Leonardo, René Descartes (1596–1650) also believed in seawater circulating through a network of subterranean channels, ultimately reaching large caverns below the mountains. There the water was supposed to evaporate due to the heat present inside the earth, while during the evaporation process the salt was left behind and the water vapour was assumed to condense when reaching the colder air surrounding the tops of the mountains. The condensed water would ultimately emerge as streamflow at the tops of the mountains. This theory remained dominant for almost two centuries.

The concept of subterranean channel networks sustaining flows of water between the sea and mountains was also defended by Athanasius Kircher (1602–1680). He thus addressed the same problem as his predecessors, i.e. the rising of waters to higher altitudes inside the mountains.

The concept of subterranean networks of flowing waters was finally rejected by John Ray (1627–1705) during the 17th century. He correctly believed that the sun attracted "*vapours*" from the earth and the sea, the wind driving those vapours from the sea towards the land, where they ultimately fell as rain. The rivers thus derived their water supply from rainfall events, while the flowing waters returned to the sea, thus closing the water cycle.

Since the end of the 18th century other tremendously important discoveries have been made in the fields of hydrology and hydraulics. However, It is not the scope of this volume to develop the advances in modern hydrology, but to compare Leonardo's ideas and concepts with our current state of knowledge.

By writing this book we want to give credit to Leonardo's unprecedented contribution to the science of hydrology, which is probably poorly known to the hydrological community. We also want to demonstrate, through his writings, the struggle he fought in his mind to bring things together towards a unifying principle: a general view of the world. However, this book can by no means give a complete view of Leonardo's work on the hydrological cycle, for many of his notes have been lost since his death. This often makes it difficult to identify both the gradual changes in his perception of certain hydrological processes, and his ultimate convictions. The confrontation of his theories and concepts about the hydrological cycle with the current state of knowledge, which we present in the following chapters, should essentially be seen as a tribute to the work of a man who lived 500 years ago and who became one of the greatest universal geniuses in the history of mankind. By his continued efforts to validate his hypotheses with experiments, he can certainly be regarded as one of the first true scientists coming out of the dark Middle Ages. This is nicely illustrated by one of his notes about "***the errors of those who depend on practice without science***", where he states that: "***He who falls in love with practice without science is like a sailor who enters a ship without a helm or a compass, and who never can be certain of where he is going***" (Codex Paris, G).

2   LEONARDO'S "TREATISE ON WATER"

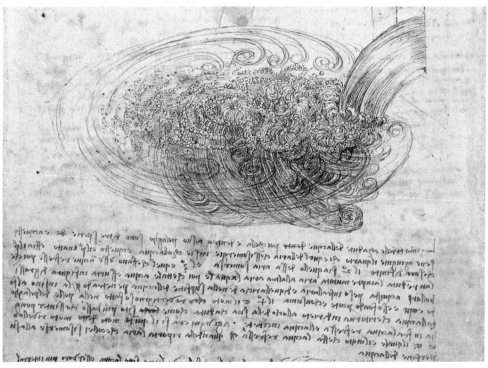

Studies of flowing water, detail (c. 1509–1511). Note Leonardo's typical back-to-front (mirror) writing.

The Royal Collection © 2008 Her Majesty Queen Elizabeth II (RL 12660v).

2.1 LEONARDO AND THE "SCIENCE OF WATER"

Leonardo's interest in water may have been inspired, at least in part, by the impressive force it sometimes displays. Leonardo described floods as the most terrifying disasters threatening humans and their buildings (Codex Atlanticus). He considered their destructive power so great, that the implementation of protective measures for human life and for structures was almost impossible. In his description of the utility of the "*science of water*" (Codex Atlanticus), Leonardo referred to large cities located close to major rivers that were heavily damaged or destroyed by devastating floods. It is interesting to note that, to this very day, of all natural disasters, floods are still the events that cause the greatest loss of lives, goods and infrastructure, all around the world.

Fig. 2.1 Representation of natural phenomena, detail (c. 1511–1512). The Royal Collection © 2008 Her Majesty Queen Elizabeth II (RL 12388r).

According to Leonardo, the "*science of water*" (Codex Atlanticus) would ultimately help to provide the means to effectively protect humans from major floods. From his point of view, flood disasters were due to the under-dimensioning of river valleys in comparison to the amount of water that travelled through them. Protective measures against floods thus had to focus on hydraulic measures to reduce the speed and the destructive force of flowing waters. In addition to his motivation to protect human life and infrastructure, Leonardo also knew that the re-arrangement of rivers and the building of canal networks was a prerequisite for the economic development of cities and countries. In his writings many examples of such studies can be found, covering engineering as well as financial issues, and even cost-benefit analyses. Many of Leonardo's writings are on hydraulics, dealing with the description and interpretation of water in motion.

In his notes, Leonardo demonstrates that he was well aware of the influence of anthropogenic activities on soil erosion caused by flowing waters. Indeed he wrote that "*the sediments of the rivers are more abundant in areas that are densely populated than in countries that are sparsely inhabited*" (Codex Atlanticus). Leonardo related this to the fact that "*in those* (densely populated) *places the mountains and hillslopes are tilled and the rain sweeps away more easily a ploughed soil than a soil that is hard and covered by vegetation*" (Codex Atlanticus). From these statements we can see that Leonardo had identified soil erosion as a significant issue, just as the loss of fertile land through erosion processes is a major concern for modern agriculture.

It is fascinating to see how Leonardo had a dual interest in the "*science of water*", driven by both scientific curiosity and engineering needs; a duality embodied by many modern hydrologists (Beven, 2006).

In his notes, Leonardo provided the structure that is required for a scientific study of the water cycle: "*First you shall make a book describing places occupied by freshwaters, and the second by salt waters, and the third, how by the disappearance of these, our parts of the world were made lighter and in consequence more remote from the centre of the world*". In the above structure most aspects of his theory on the water cycle can be found, with a distinct separation between the freshwater flowing from the springs on the surface of the earth to the salt waters of the oceans. During its downstream flow, surface water would erode the soils and rocks through which it flows and would thus progressively lighten the continents. Due to this weight loss, these parts of the earth would continuously rise and create new relief. As Leonardo stated, the fossils that can be found in mountainous areas are the witnesses of past morphological changes of the earth's surface. Indeed, he wrote that: "*the shells, oysters, and other similar animals, which originate in sea-mud, bear witness to the changes of the earth round the centre of our elements*" (Codex Paris, E).

Leonardo provided a series of principles and rules related to science. He considered science as "*the observation of things possible, whether present or past; prediction is the knowledge of things which may come to pass*" (Codex Trivulzio). He recognised the importance of experience in the verification process of a hypothesis, writing that "*experience, the interpreter between nature and humans, teaches how nature acts upon mortals; and being constrained by laws cannot act otherwise than through reason, which is its helm, requiring her to act*" (Codex Atlanticus).

He was also well aware of the risk related to drawing the wrong conclusions from scientific experiments. In this respect he wrote that "*experience does not err; only your judgments err by expecting from her what is not in her power. Men wrongly complain of experience; with great abuse they accuse her of leading them astray but they set experience aside, turning from it with complaints as to our ignorance causing us to be carried away by vain and foolish desires to promise ourselves, in her name, things that are not in her power; saying that she is fallacious. Men are unjust in complaining of innocent experience, constantly accusing her of error and of false evidence*" (Codex Atlanticus).

Leonardo intended to record the knowledge he had gained about water in a *Treatise on Water*. Many of his ideas and concepts can still be taken from his writings, even though thousands of pages of his work seem to definitely be lost. The fact that there is no clear structure to his writings represents an additional difficulty that we will try to at least partially balance by investigating the structure Leonardo intended to give to his *Treatise on Water*.

2.2 LEONARDO'S UNFINISHED "TREATISE ON WATER"

For Leonardo, everything had to do with everything, and this probably motivated his interest in an impressively wide variety of fields; fields which, to most of us, may seem remote and very different from each other. His interests ranged from the arts to science and engineering. His studies, discoveries, and intuitions in all these fields often benefited from each other. In each he introduced new ideas, often in stark contrast with the tradition, and his discoveries preceded by several centuries those of formal present-day science.

In the field of natural science, he dedicated his attention to: *Geology*, investigating the processes of erosion and sedimentation and the origin of fossils; *Botany*, being the first to deduce the age of trees by observing the number of concentric circles of their sections; *Hydraulics*, establishing before any other, the principle of conservation of mass in a stream channel and understanding the effect of flow on river morphology (see Chapter 2 frontispiece and Fig. 2.2); *Aerodynamics*, where by studying the flight of birds he managed to explain how the interaction of forces between solids and air vary with relative velocity; and *Hydrology*, trying to derive an explanation for the hydrological cycle.

The body of his work in arts and science that is still available today, encompasses a large amount of incomplete works. This is both because he was continuously distracted by new research topics, and perhaps because he did not consider his work merely as a job that had to be finished. He was probably well aware of this since in his writings he often used words like "***remember***", "***describe***", "***don't forget***", followed by the things that he planned to realize in the future.

Leonardo's "*Treatise on Water*" was to cover a whole variety of topics that would explain, in an ordered and systematic way, aspects related to the water cycle as well as the engineering structures related to water. One of the outlines Leonardo prepared for this book is as follows (Codex Leicester):

Chapter 1: "***On water in itself***"
Chapter 2: "***On the sea***"
Chapter 3: "***On subterranean rivers***"
Chapter 4: "***On rivers***"
Chapter 5: "***On the nature of the abyss***"
Chapter 6: "***On the obstacles***"
Chapter 7: "***On gravels***"
Chapter 8: "***On the surface of water***"
Chapter 9: "***On the things placed therein***"

Chapter 10: "*On the repairing of rivers*"
Chapter 11: "*On conduits*"
Chapter 12: "*On canals*"
Chapter 13: "*On machines driven by water*"
Chapter 14: "*On how to rise water*"
Chapter 15: "*On the destructions caused by water*"

Fig. 2.2 Studies of flowing water, detail (c. 1510–1915). The Royal Collection © 2008 Her Majesty Queen Elizabeth II (RL 12579r).

Despite multiple attempts, Leonardo never completed his *"Treatise on Water"*. Indeed, there are many indications that he regularly tried to organise his notes on water. At one stage he obviously intended to write several books, which were to *"contain in the beginning: of the nature of water itself in its motions; the others treat of the effects of its currents, which change the world in its centre and its shape"*.

In Leonardo's writings we can also find attempts to develop his ideas concerning the *"effects of rivers"*, as well as what hydraulic measures might be taken to reduce negative effects. For example, he intended to write chapters about (Codex Leicester):

- *guarding against the impetus of rivers so that towns may not be damaged by them*
- *training of rivers so as to preserve their banks*
- *how to make or to repair the foundations for bridges over the rivers*
- *the repairs which ought to be made in walls and banks of rivers where the water strikes them*
- *diverting rivers from places where they do mischief*
- *controlling rivers so that the little beginnings of mischief, caused by them, may not increase*
- *on the various motions of waters in their rivers*
- *on the windings and meanderings of the currents of rivers*
- *the soils which absorb water in canals and of repairing them*, or
- *on creating currents for rivers, which quit their beds, and for rivers choked with soil*.

Although many of Leonardo's writings have been lost over time, those that are available allow reconstruction of some of his ideas. Those writings, now collected in a variety of codices, were originally just a collection of papers on which he noted down his ideas. His style is sometimes extremely concise and difficult to interpret, and sometimes very clear and detailed. Sometimes he writes to himself, just to remember something, while on other occasions he imagines an opponent whom he has to convince. Often the topics treated are incomplete and range from one problem to a completely different one. However, the patchwork that we can make by assembling all his texts dealing with water, is a demonstration of his great vision on what the earth is made of, and on the processes that shape our world.

Most of the questions he raised have now been answered and are part of common knowledge in hydrological science. Some of his ideas are now known to be wrong, and sometimes, with the benefit of hindsight, it seems incredible that he did not understand things that appear so obvious to us now. Despite that, there is probably still a lot we can learn from him. There are ideas that are still modern and valid today, and which go beyond the mere historic interest in a man who, 500 years ago, tried to deal with the complicated subject of hydrology.

Leonardo used a scientific approach. In the development of his theories, he aimed to learn from experience, stating that a major task of the scientist is to reduce the discrepancy between theory and practice to the minimum possible. Leonardo writes: *"Remember, when you speak about water, to show first the experience* (empirical evidence) *and then the interpretation"* (Codex Paris, H), and most of his reasoning in every field is advanced using empirical evidence.

This made him a scientist far ahead of his time, who was also able to admit that he was wrong, rejecting a hypothesis in favour of a new one. How different this is to how most hydrologists work today! While the scientific approach of hypothesis testing is well established, we are often too much conditioned by conventional models to realise that something is wrong.

Leonardo had a zeal for confronting traditional beliefs and challenging established truths. He always tried to think independently, developing his own theories without following the established authority and tradition too much. His research developed outside the academic training and common curricula of the science of his times. Nowadays, we often put ourselves in a situation of working in a context where we know all the boundary conditions, and where we know beforehand the objectives of our work.

Leonardo attempted to relate things to each other, so making them part of an integrated holistic theory on the behaviour of our world. We as hydrologists rarely make this effort. Even if we have evidence of recurring patterns, preferred distributions or states of the natural system, e.g. arrangements of soil depths, soil moisture, vegetation, or configurations of preferential flow, we rarely try to combine them together into a holistic view of the system; we prefer to analyse them within the boundaries of our expertise.

2.3 THE CONCEPTUAL FOUNDATIONS OF LEONARDO'S "TREATISE ON WATER"

As a whole, Leonardo's *Treatise on water* was intended to deal with "***the nature of water itself and its motions***" (Codex Paris, E). Although he never completed the book, he provided a number of indications and definitions relevant to each of the chapters he had planned.

In his notes, Leonardo gave definitions of what can be called his laws on water, which he considered prerequisites to the understanding of the water cycle. Those laws are difficult to understand at first sight, but they become clearer when we compare them with his general ideas that emerge from reading the whole body of his writings (Codex Arundel; Codex Forster; Codex Paris, A):

1. *The lowest point of elements is their centre. The common centre is that which is equally remote from the extremes or from the opposite weights of the elements, and nothing can be imagined that is less than the centre, and this is what is called the point by mathematicians.*
2. *Lowness and highness. The lower thing is that which is closer to the centre of the elements, therefore the higher thing is that which is more remote from its centre.*
3. *Gravity. Gravity is an accident* (a phenomenon) *that is generated because of violence of an element thrown or pushed into another.*
4. *Lightness. Lightness is that thing that is less heavy than another, because in elements everything is heavy, but it can be more or less heavy than something else.*
5. *Elements. No part of an element, if not moving, has gravity or lightness in its own element.*

6. *Water does not acquire gravity if it is not connected with an element or another liquid that is lighter and lower than itself.*

Regarding water in motion, Leonardo wrote that: "*water does not move if it does not descend. Therefore, if it moves, it descends, and this happens because of the 6th* (law)*. And if you say that in the reflected motion that it takes after its descent it rises high because of itself, this is not true, it is an accident* (event)*, as it is proven by the 3rd* (law)*. This accident is not part of water, and has no similitude with it, and it does not remain with it, because it has a short life*" (Codex Leicester). Here Leonardo talks about a hydraulic jump, which he considers an event where the elements are temporarily out of balance (i.e. the acceleration terms in the equation of motion) and where gravity will bring the elements back to where they belong. In the same context, he also stated that "*water does not stop if it is not contained with equal height*". Hence he states that there is no flow without slope and still water has no slope.

The first law becomes clear when we realise that, for Leonardo, the four elements earth, water, air and fire are concentric spheres that surround each other. The earth occupies the first sphere, it is surrounded by water, then by air, and then by fire. All the four elements tend to remain in their respective spheres, and when they are separated they will try to go back to where they belong. The centre of these spheres is what we would call the point of lowest potential. Since a point has no dimensions, Leonardo writes that nothing can be considered to be smaller than a point. If we neglect the "*element*" of fire, then this theory makes sense. It states that, if given the opportunity, the heaviest elements will tend to be closest to the earth's centre and that they will order themselves according to their density: rocks nearest to the centre of the earth, then water and then air. If water is thrown into air, which is lighter, it will tend to go back to its own location between rock and air. In fact this concept agrees with the modern concept of entropy, whereby substances order themselves into their most probable state. An event can mix them up, but eventually they will rearrange themselves to achieve maximum entropy, which is a steady state.

What seems strangest to us now, is that fire is seen as an element; an element residing in the outer sphere. The experience underlying this concept may be that a flame always tends to go upward and that lightning (an event) appears to come from the outer sphere to the inner spheres, after which the situation stabilises again. Also, observations of the Northern Lights (aurora borealis), which occur in summer at high latitudes, might support this theory. Still, even within the framework of this theory, we would consider fire an event, rather than an element.

Leonardo writes that water has two centres, a global centre, as expressed above, and a local centre, which is the centre of water drops. The universal centre of the water sphere refers to all waters not in motion, i.e. "*canals, ditches, ponds, fountains, wells, dead rivers, lakes, stagnant pools and seas, which, although they are at various levels, have each in itself the limits of their superficies equally distant from the centre of the earth, such as lakes placed at the tops of high mountains*" (Codex Leicester). He relates the roundness of the water drop to the roundness of the element in general, and he notes that the centre of the water drop tends to move outside the water drop towards the common centre of water as the weight of the drop increases.

Fig. 2.3 Studies of flowing water, detail (c. 1509–1511). The Royal Collection © 2008 Her Majesty Queen Elizabeth II (RL 12660v).

In order to understand the third and fourth laws (on gravity and lightness), it has to be recalled that for Leonardo gravity does not have the same meaning as we give it today. For Leonardo, gravity is not an intrinsic property of things; it is an event, something that elements acquire when they are removed from their place, and "***has life as long as an element is struggling to go back to its place***" (Codex Paris, B). By "***gravity***" (Codex Paris, F), Leonardo means that if an element is heavier than the upward force of buoyancy (in another element), it sinks. With "***lightness***" (Codex Paris, F), he indicates that if the weight is less than the upward force of buoyancy, it moves up. If they are equal, the element appears weightless. This is, of course, the case with water in water and air in air, but also with a diver who, by controlling his buoyancy, hovers at a fixed depth. So, in fact, Leonardo's "***gravity***" is the difference between the weight of an element (the force of gravity, as we define it, acting on an element) and the buoyancy (the upward force that an element experiences when submersed in another element). Negative "***gravity***" is "***lightness***". With these terms Leonardo indicates that an element is not in equilibrium, but he also assigns to it the meaning of a vital force acting on inanimate bodies, making them similar to living ones.

As a consequence, Leonardo stated that within its own sphere, an element has no weight (the fifth law). Although we know that this concept is wrong, it is, however, both practical and understandable. Under hydrostatic conditions, where there is vertical equilibrium between weight and buoyancy, there is no vertical movement of an element within its own sphere; hence, it appears weightless. To move an element out of its sphere, an additional force is needed (an event) that carries the element into the sphere surrounding it. But this is temporary, as gravity will bring it back to its own sphere, eventually. The meaning of the fifth law becomes clearer after reading the following: "*Water that has no motion does not exercise weight on its bottom, as it is possible to see for the very fine herbs waving in the water, and for the very light mud on the bottom of marshes, which is almost as light as the water. Hence, if water weighted on it, this would become like stone, and showing that this does not happen, it can be demonstrated that water does not have weight*" (Codex Arundel). In other words, Leonardo argues that because water does not crush the aquatic plants or compress the fine sediment at the bottom of a pond or stream, it does not have any weight when it is in its own sphere.

In addition to these definitions of the main laws that water obeys, Leonardo also provided concepts and ideas that can be more specifically related to the chapters he had planned for his book on water.

Chapter 1. "*On water in itself*"

Leonardo knew that our world is made of earth and water, having the shape of a sphere, and considered the centre of the sphere of water to be the true centre of the globe. In this context, he correctly wrote that "*if you want to find the centre of the element of the earth, this is placed at a point equidistant from the surface of the ocean, and not equidistant from the surface of the earth; for it is evident that this globe of earth has nowhere any perfect rotundity, excepting in places where the sea is, or marshes or other still waters. And every part of the earth that rises above the water is farther from the centre*" (Codex Paris, A).

As will be shown in Chapter 5 of this publication, Leonardo frequently questioned himself about the forces that generate the motion of water on and inside the earth. From his writings it looks as if he had a concept of the water cycle that relied on two different heat sources driving the planetary water cycle. He considered that there is a "*vital heat, and where vital heat is, there is movement of vapour*" (Codex Paris, A). This heat was supposed to be generated by both the sun and the "*element of fire*", which Leonardo thought to exist beyond the outer limits of the atmosphere. His writings prove that he clearly understood that this heat evaporates water, leading in the end to the formation of clouds in the sky, since "*heat and moisture cannot exist with cold and dryness*" as found at higher altitudes in the atmosphere. Leonardo stated that "*if the heat of the sun is added to the power of the element of fire, the clouds are drawn up higher still and find a greater degree of cold, in which they form ice and fall in storms of hail*" (Codex Paris, A). Like many of his contemporaries, Leonardo did not appreciate that evaporation and precipitation of water are the main driving forces of the water cycle, and as such feed the rivers. As we explain in Chapter 5, he wrote in his notes that the "*element of fire*" causes a rise of water "*from the foot of the mountains, and leads and holds them within the summits of the mountains, and these, finding some fissure, issue continuously and cause rivers*" (Codex Paris, A).

Chapter 2. "*On the sea*"

In his notes on the oceans, Leonardo insists on his belief that "*the ocean does not penetrate under the earth*" (Codex Paris, G). He gives two examples of observations that he made proving this assumption. First he refers to "*the many and various springs of fresh water which, in many parts of the ocean make their way up from the bottom to the surface*" (Codex Paris, G). As a second example, Leonardo describes "*wells dug beyond the distance of a mile from the said ocean, which fill with fresh water; and this happens because the fresh water is lighter than salt water and consequently more penetrating*" (Codex Paris, G).

Leonardo interrogated himself about the origin of the salt in the sea water, writing his discussion in the manner of a rebuff to Pliny's theory regarding the saltiness of the sea. First he cites Pliny: "*the water of the sea is salty because the heat of the sun dries up the moisture and drinks it up; and this gives to the wide stretching sea the flavour of salt*" (Codex Paris, G). Leonardo rejects this argument, arguing that if Pliny was right, then his theory would have to apply equally well to lakes, pools and marshes, i.e. their waters would also have to be salty, which is of course not the case. He concludes that "*the saltiness of the sea must proceed from the many springs of water which, as they penetrate into the earth, find mines of salt and these they dissolve in part, and carry with them to the ocean and the other seas, whereas the clouds, the begetters of rivers, never carry it up*" (Codex Paris, G). According to Leonardo, the fact that the salt is carried to the sea by the rivers explains that the "*waters of the salt sea are fresh at the greatest depths*". Furthermore, he developed his theory, writing that "*the sea would be saltier in our times than ever it was at any time; and if the adversary were to say that in infinite time the sea would dry up or congeal into salt, to this I answer that this salt is restored to the earth by the setting free of that part of the earth which rises out of the sea with the salt it has acquired, and the rivers return it to the earth under the sea*" (Codex Paris, G).

As it will be shown in Chapter 5, Leonardo also dedicated many of his efforts to the understanding of the ebb and flow of the tide. He questioned himself on "*whether flood and ebb are caused by the moon or the sun, or are the breathing of this terrestrial machine. That the flood and ebb are different in different countries and seas*" (Codex Leicester).

Chapter 3. "*On subterranean rivers*"

Leonardo was convinced that "*very large rivers flow under ground*" (Codex Atlanticus), forming a network of subterranean conduits where a large part of the planetary water cycle was supposed to take place. Basically, he thought that water moved from the bottom of the seas to the top of the mountains through numerous branching conduits. The explanations he gave for this upward movement changed somewhat over time. Leonardo often relied on what he believed to be similarities between living bodies and the earth, to explain the cause of the upward movement of water in the subterranean conduits. He thus wrote that "*the same cause which stirs the humours in every species of animal body and by which every injury is repaired, also moves the waters from the utmost depth of the sea to the greatest heights*" (Codex Arundel).

Despite his numerous and well known studies on the anatomy of man and animals, Leonardo had not immediately perceived the role of the heart as a pump for moving blood through living bodies. For a long time he considered heat as being at the origin of the movement of blood, saying that *"the natural heat of the blood in the veins keeps it in the head of man, for when the man is dead the cold blood sinks to the lower parts, and when the sun is hot on the head of a man the blood increases and rises so much, with other humours, that by pressure in the veins pains in the head are often caused"* (Codex Paris, A). Leonardo extended this assumption to the earth, stating that *"in the same way veins ramify through the body of the earth, and by the natural heat which is distributed throughout the containing body, the water is raised through the veins to the tops of mountains. ... Again, the heat of the element of fire and, by day, the heat of the sun, have power to draw forth the moisture of the low parts of the mountains and to draw them up, in the same way as it draws the clouds and collects their moisture from the bed of the sea"* (Codex Paris, A).

Well aware of the fact that the water cycle is never-ending, Leonardo gave additional detail to his concept, writing that water *"rises from the utmost depths of the sea to the highest tops of the mountains, and flowing from the opened veins returns to the low seas; then once more, and with extreme swiftness, it mounts again and returns by the same descent, thus rising from the inside to the outside, and going round from the lowest to the highest, from whence it rushes down in a natural course"* (Codex Arundel).

According to Leonardo, there were thus two complementary forces causing the water cycle to be endless: the vital heat for the rise of water and gravity for the downward flow of water via the rivers.

Chapter 4. *"On rivers"*

For Leonardo, the focus of this chapter was to be the *"way in which the sources of rivers are fed"* (Codex Leicester). Here again, he wrote about the similarities between living bodies and the body of the earth. Essentially, he believed that the subterranean rivers were progressively eroding the conduits, ultimately causing those that are close to the earth's surface to collapse: *"**The subterranean channels of waters, like those which exist between the air and the earth, are those which unceasingly wear away and deepen the beds of their currents**"* (Codex Leicester). Leonardo argued that those who said that the rivers are fed by rain and melting snow were wrong, referring to *"the rivers which originate in the torrid countries of Africa, where it never rains – and still less snows – because the intense heat always melts into air all the clouds which are borne thither by the winds"* (Codex Leicester). Here his knowledge of Africa apparently stopped at the Sahara or the Savannah lying south of it. He apparently never knew of the wet tropical areas lying south of the Sudan.

Leonardo also studied the erosion process in relation to waters flowing downslope, driven by gravity. This part of his work is analysed further here in Chapter 4.

The writings that are known to us today and deal with issues related to the study of water, do not clearly relate to all the chapters that Leonardo had initially planned for his *"Treatise on Water"*. Indeed, it is very difficult to arrange his notes, especially since it is not always clear when and why he changed some of his ideas and concepts related to the water cycle. Rather than strictly relying on Leonardo's structure for his book, we have put his writings into our own thematic framework. In the following chapters we will compare his concepts of the water cycle and water in motion to present day knowledge, looking successively at atmospheric processes, the geological processes that shaped the earth, the hydrological cycle and finally, river hydraulics.

3 THE ATMOSPHERE

Heavy storm over riders and trees, detail (c. 1514).
The Royal Collection © 2008 Her Majesty Queen Elizabeth II (RL 12376r).

3.1 CONSTITUENTS AND STRUCTURE OF THE ATMOSPHERE

The atmosphere is a thin layer of gases and dust particles surrounding the earth. It can be divided into three layers, each having its own temperature characteristics, namely the troposphere, the mesosphere and the thermosphere. It consists of a mixture of gases containing 78% nitrogen, 21% oxygen, 1% argon, less than 1% carbon dioxide, and tiny proportions of neon, krypton, helium, methane, xenon, hydrogen and ozone, as well as water vapour, the latter varying between 0 and 4%. Oxygen and nitrogen are not very efficient in absorbing radiation, whereas the small quantities of ozone, nitrous oxide, water vapour, methane and carbon dioxide have a significant effect on atmospheric absorption of radiant energy, contributing to the global greenhouse effect (Whittow, 2000).

Water vapour corresponds to the non-visible state of water in the atmosphere and its presence is a prerequisite for any form of precipitation, while at the same time it plays an essential role in the earth's heat balance. Together with carbon dioxide, it forms a global shield that prevents excessive heat loss via terrestrial radiation.

The need to understand the colour of the atmosphere, its variation in time and space, and its effect on visible objects certainly motivated Leonardo's interest in the atmosphere. This interest was related to the art of painting, where his scientific findings took shape and became expression. Leonardo understood that colour is the effect of physical processes, which require interpretation and explanation. Leonardo's interest in meteorological phenomena, however, was also very strongly related to his desire to develop a flying machine. Whilst studying bird flight and developing concepts of artificial wings, Leonardo anticipated the importance of stable meteorological conditions for safe flight conditions. Using words that are different to ours, he gave an explanation of atmospheric processes that closely reflects our current understanding. In his writings he described the effect of water vapour on the brightness of the atmosphere, as well as its role on the formation of clouds. He understood that the atmosphere consisted of an infinite number of small gas molecules. While questioning himself on the constituents of the atmosphere and their role on the blue colour of the sky, Leonardo wrote that "*the blueness we see in the atmosphere is not an intrinsic colour, but is caused by warm vapour evaporated in minute and inert particles on which the solar rays fall, rendering them luminous against the infinite darkness of the sphere of fire which lies beyond it and surrounds it*" (Codex Leicester).

He supports this conclusion by noting that "*if the atmosphere had blue as its natural colour, the deeper the layer between the eye and the sphere of the fire, the stronger the colour would be, as it is possible to see in blue glass and sapphires, which become darker as they are bigger. The air behaves in the opposite way, ... it is whiter close to the horizon, ... and darker when less air is interposed between the eye and the sphere of fire*" (Codex Leicester).

Leonardo also observed that: "*when overcharged with moisture, the atmosphere appears white, while the small amount of heated moisture makes it dark, of a dark blue colour*" (Codex Leicester).

He described how the atmosphere, through its interaction with sunrays, causes the stars in the sky to be invisible by day: "*The stars are visible by night and not by day, because we are beneath the dense atmosphere, which is full of innumerable particles of moisture, each of which independently, when the rays of the sun fall upon it, reflects a radiance, and so these countless bright particles conceal the stars; and if it were not for this atmosphere the sky would always display the stars against its darkness*" (Codex Leicester).

These writings show a partially correct understanding of the scattering of the sunlight by the atmosphere. He describes the composition of the atmosphere itself well, stating that it is formed of atoms (although usually mainly referring to water vapour), when attributing the brightness of the sky to the illumination of the atmosphere's constituents by the sun: "*the blueness we see in the atmosphere is not intrinsic colour, but is caused by warm vapour evaporated in minute and insensible atoms on which the solar rays fall, rendering them luminous against the infinite darkness of the fiery sphere which lies beyond and includes it*". He was also right in assuming that in a humid air the sky is pale blue, but the reason he mentioned for the blue colour of the sky, namely the darkness that is hidden behind the illuminated atmosphere, is false. Here it must be stated that Leonardo was in no way referring to the atom as a basic unit of matter known in modern physics and which was defined as such only in the 17th and 18th centuries by Robert Boyle (atoms as corpuscles) and Antoine Lavoisier (atoms as elements). It is very likely that he relied on the so-called theory of atomism proposed by the ancient Greek philosophers Leucippus (5th century BC) and Democritus (*c.* 460–370 BC), defining the "atomos" (i.e. uncuttable) as the smallest indivisible particle of matter.

In fact, the scattering of the sunlight by the molecules of atmospheric gases mostly affects the short wavelengths of light making the sky appear blue. Since there are always some dust particles in the air, longer wavelengths also contribute to the sky's colour. These cause a mixture of other colours, such as green and red, with blue ultimately making the sky look paler (Estienne & Godard, 1990). This can be observed clearly in industrial areas, where the atmosphere is usually much paler than in rural areas, where a deep blue sky can quite frequently be observed. Moreover, in humid air masses condensation nuclei absorb water vapour from the air, causing the sky to appear paler than in dry air masses.

Leonardo also had a precise perception of the decrease of atmospheric pressure with altitude. The gaseous envelope that surrounds the earth can be compared to a compressible fluid. The pressure decreases with altitude according to a logarithmic law, with the constituents of the atmosphere having a greater density in the vicinity of the earth's surface. Leonardo illustrated the decrease of atmospheric density with altitude, describing how, when standing on the summit of high mountains in the Alps, he "*saw above himself the dark sky, and the sun as it fell on the mountain was far brighter here than in the plains below, because a smaller extent of atmosphere lay between the summit of the mountain and the sun*" (Codex Leicester).

3.2 METEOROLOGICAL PHENOMENA

In addition to his descriptions of the constituents and structure of the atmosphere, Leonardo provided explanations, sometimes incomplete, of various meteorological phenomena. He tried to understand the processes behind the formation of clouds and their migration in the sky. In this context, his objective was to "*describe how the clouds are formed and how they dissolve, and what causes the rise of vapour*". As discussed in Chapter 5, for several decades he studied the origin of the water cycle and the various processes driving it. Water vapour, stemming from evaporation of liquid water in response to heating by sunlight, is the major component of the atmospheric part of the water cycle. Today, we know that meteorological phenomena are part of a never ending cycle of redistribution of different forms of energy on our planet.

Regarding evaporation and condensation he explained: "*One time in 24 hours the humidity excess that is in the air falls on the earth, and then it rises, drawn and sustained by the heat. Then, being abandoned by that, it falls on the earth. Such humidity is called dew during summer, and frost during winter, because it is frozen by the cold*" (Codex Paris, C).

Leonardo gave a detailed description of the formation of clouds and raindrops. Basically, he considered that the higher the moisture particles in a cloud rise, the more they are cooled due to the low temperatures that prevail at high altitudes. This cooling effect causes a progressive deceleration during the ascent phase of these particles. According to Leonardo, the moisture particles that continue to rise from the lower parts of the atmosphere progressively merge with those already in place and slow down under the cooling effect. Finally, those particles that are "*acquiring weight, also acquire velocity*" (Codex Atlanticus) during their descent through the clouds. The World Meteorological Organization (WMO) defines the clouds as a "*visible aggregate of minute particles of water or ice, or both, in free air*". A drop radius in the range of 0.1 to 0.25 mm corresponds to drizzle, while larger drops are defined as rain. Drop formation is due to a process called nucleation, where colliding aerosols and water vapour molecules ultimately combine (Strangeways, 2007). Drop interception leads to an increasing concentration of large raindrops. Being subject to gravity, they end up falling to the ground as precipitation.

In his reflections on the generation of thunderstorms, Leonardo observed: "*like the sun that hits concave mirrors and that is reflected by them in a pyramidal shape, and the part of the pyramid is proportionally warmer than its base as much as it is smaller than that, so acts humidity that condenses: the warm vapour that is mixed into it becomes more powerful as much as it comes together, and the smaller is the place that contains it, the more heat it generates. Therefore often it lights up, generating a burst in the cloud; hence the cloud breaks in lightning and thunders. The small grains of water, as the cloud is restricted by cold, join together, and fall because of their weight*" (Codex Arundel).

Evidently Leonardo had a clear view of the processes of evaporation, condensation, and rainfall generation. He understood the balance between heat and humidity contained in the atmosphere, which drives the processes of evaporation and condensation. However, his appreciation of the process of thunderstorm generation did not come very close to our present knowledge. Leonardo identified the importance of high moisture and temperatures in the lower atmosphere as a prerequisite for the formation of thunderstorms, but did not know about the other two main causes, namely a significant decrease in air temperature with increasing altitude, as well as a strong lifting force acting on the air mass (e.g. warming of the earth's surface, orographic lifting) (Beltrando & Chémery, 1995).

Leonardo also studied the origins of the colours of the rainbow and set up a number of experiments to identify them. His explanations focused on the role of the sun and the eyes of the observer. One experiment was intended to prove that the sun does not create the colours: *"the colours of the rainbow are not produced by the sun, for they occur in many ways without sunshine; as may be seen by holding a glass of water up to the eye; when, in the glass – where there are those minute bubbles always seen in coarse glass – each bubble, even though the sun does not fall on it, will produce on one side all the colours of the rainbow; as you may see by placing the glass between the daylight and your eye in such a way as that it is close to the eye, while on one side the glass admits the (diffused) light of the atmosphere, and on the other side the shadow of the wall on one side of the window; either left or right, it matters not which. Then, by turning the glass round you will see these colours all round the bubbles in the glass"* (Windsor Drawings).

He also attempted to demonstrate that the rainbow's colours were not due to some kind of interaction by the eyes: *"In the experiment just described, the eye would seem to have some share in the colours of the rainbow, since these bubbles in the glass do not display the colours except through the medium of the eye. But, if you place the glass full of water on the window sill, in such a position as that the outer side is exposed to the sun's rays, you will see the same colours produced in the spot of light thrown through the glass and upon the floor, in a dark place, below the window; and as the eye is not here concerned in it, we may evidently, and with certainty pronounce that the eye has no share in producing them"* (Windsor Drawings).

Having conducted the above experiments, Leonardo concluded that rainbows are actually the result of an interaction between the rain, the sun and the observer's eyes: *"the bow in itself is not in the rain nor in the eye that sees it; though it is generated by the rain, the sun, and the eye. The rainbow is always seen by the eye that is between the rain and the body of the sun; hence if the sun is in the East and the rain is in the West it will appear on the rain in the West"* (Codex Paris, E).'

These writings demonstrate that Leonardo's explanation of the rainbow comes close to the explanation prevailing today. A rainbow is an optical phenomenon forming a luminous arc across the sky comprising all spectral colours with red on the outer side and violet on the inner. It is caused by refraction and internal reflection of sunlight through falling raindrops, when the observer stands with his back to the sun. The larger the drops, the more spectacular the colours.

Leonardo was also interested in storms and tornados; he describes their effects with particular attention and detail. In his writings on the *"motion of air"*, he observes that *"the eddies of wind .... strike upon the waters and scoop them out in a great hollow, whirl the water into the air in the form of a column, and of the colour of a cloud. And I saw this thing happen on a sand bank in the Arno, where the sand was hollowed out to a greater depth than the stature of a man; and with it the gravel was whirled round and flung about over a great space; it appeared in the air in the form of a great bell-tower; and the top spread like the branches of a pine tree, and then it bent at the contact of the direct wind, which passed over from the mountains"* (Codex Leicester; see also Fig. 3.1).

While making this detailed description of a whirlwind, Leonardo also tried to work out what could be the cause of this and other meteorological phenomena. At a later stage of his life, he was finally convinced that it is the heat of the sun and especially that of the *"element of fire"* (Codex Atlanticus) that actually causes the air to move. Indeed, he wrote that *"the element of fire acts upon a wave of air in the same way as the air does on water, or as water does on a mass of sand – that is earth;*

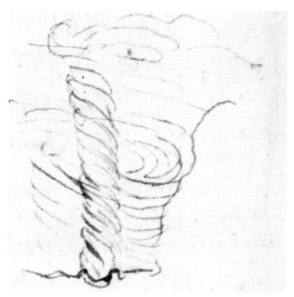

Fig. 3.1 Sketch of a whirlwind. Codex Leicester. © Seth Joel/CORBIS.

and their motions are in the same proportions as those of the drivers acting upon them". The latter sentence may be interpreted as: *action equals reaction*.

In the next section, we describe how Leonardo developed his own instruments to observe meteorological phenomena and to understand the physics underlying meteorological processes.

3.3   MEASURING DEVICES

In the development of his theories, Leonardo always looked for empirical evidence. His principle "measuring devices" were his own senses, and most of his arguments are based on what one can see, smell, taste, hear or touch. However, he understood the value of quantitative data in providing additional evidence on how things might work, and he often built experiments and constructed measuring devices. The objective of measurement was for him the same as it is for us today: to understand and predict. Among the many tools that he constructed, we present here those that are more closely related to his understanding of atmospheric phenomena.

In order to measure the direction in which the wind blows, he designed an instrument as shown in Fig. 3.2. Next to the figure he wrote the legend "***to know better the direction of the winds***" (Codex Paris, H). This instrument is a sort of wind vane that swings around in the wind and shows the direction from which the wind blows.

Being also interested in the wind's speed, Leonardo designed an anemometer (Fig. 3.3). The force of the wind was determined by reading on a quadrant scale the highest point to which the vane, hinged at the top, was blown. The principle of this instrument was based on Leonardo's conviction that "***the air moves like a river and carries the clouds with it, just as running water carries all the things that float upon it***" (Codex Atlanticus).

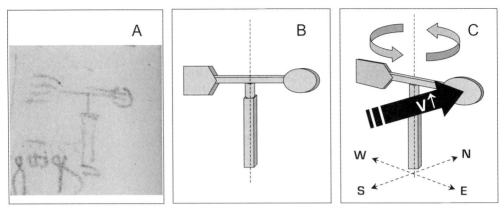

Fig. 3.2 (A) Sketch of a wind vane by Leonardo Da Vinci (c. 1493–1494). Codex Paris, Manuscript H. © RMN (Institut de France)/René-Gabriel Ojéda, Paris, bibliothèque de l'Institut. (B) Diagram of the equipment. (C) Illustration of wind acting on the equipment; V is wind velocity.

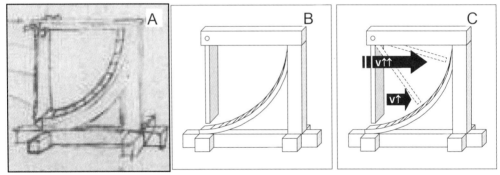

Fig. 3.3 (A) Sketch of an anemometer by Leonardo Da Vinci (c. 1500). Codex Atlanticus. Copyrights Biblioteca Ambrosiana Auth No. Int 59/08. (B) Diagram of the equipment. (C) Illustration of the wind acting on the equipment; V is wind velocity.

Leonardo went even further and considered "*measuring distance travelled per hour with the force of the wind. Here a clock is required for showing (travel) time*" (Codex Atlanticus).

Besides wind direction and wind speed, he was also interested in measuring air humidity. The devices he considered for measuring air humidity were primarily meant to show "*when the weather is breaking*" (Codex Atlanticus). One of his hygrometers consisted of a scaled balance with two containers, one containing a hygroscopic substance (cotton wool) and another containing wax (which does not absorb water). In a dry atmosphere the scale should indicate zero humidity. With increasing moisture in the air, the weight of the hygroscopic substance would increase, causing the tipping of the scale towards the container with the cotton (Fig. 3.4).

Leonardo designed another instrument for measuring the moisture content of the air. The principle of this hygrometer was identical to that above, with a hygroscopic substance changing its weight due to air humidity changes against a constant weight of wax (Fig. 3.5), the change being measured on a scale.

These instruments prove that Leonardo had a remarkable perception and understanding of the atmosphere, which was close to our current knowledge. His instruments, his data, and the value of these data to shed new light on system behaviour are similar to the way in which we try to address the key questions of our time, our limited understanding of the processes, and our need for new measurement approaches, such as formulated under the Prediction in Ungauged Basins (PUB) programme of the IAHS (International Association of Hydrological Sciences).

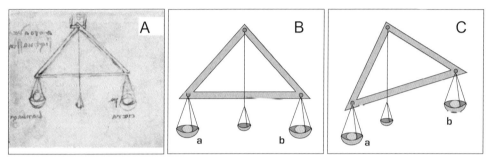

Fig. 3.4 (A) Sketch of a hygrometer by Leonardo Da Vinci (c. 1480–1486). Codex Atlanticus. Copyrights Biblioteca Ambrosiana Auth No. Int 59/08. (B) Diagram of the equipment at low atmospheric humidity; a contains a hygroscopic substance and b contains wax. (C) Diagram of how the equipment reacts to increased atmospheric humidity; the hygroscopic substance takes up water.

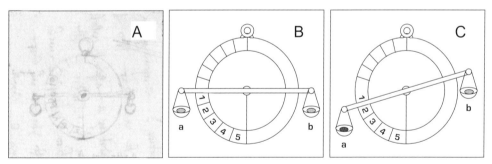

Fig. 3.5 (A) Sketch of another hygrometer by Leonardo Da Vinci (c. 1483–1486). Codex Atlanticus. Copyrights Biblioteca Ambrosiana Auth No. Int 59/08. (B) Diagram of the equipment at low atmospheric humidity; a contains a hygroscopic substance and b contains wax. (C) Diagram of how the equipment reacts to increased atmospheric humidity; the hygroscopic substance takes up water.

4 THE PHYSICAL STRUCTURE OF THE EARTH

Study of rock formations (approx. 1510–1513).
The Royal Collection © 2008 Her Majesty Queen Elizabeth II (RL 12394r).

4.1 THE EARTH AS A SUPPORT FOR THE WATER CYCLE

Leonardo spent a lot of time attempting to conceptualize the global water cycle. Today's knowledge of the water cycle is based on atmospheric, surface and subsurface transport processes of water in its gaseous, liquid and solid states. The engine driving the water cycle is solar radiation causing evaporation, condensation, cloud formation, transport of moisture over large distances, and eventually precipitation of moisture onto the earth, mostly during rainfall events. Raindrops either run-off or evaporate. The contribution to runoff is delayed in time when infiltration processes are dominating, driving the replacement of old water.

Although contemporary scientists can rely on countless sophisticated instruments designed specially for measuring numerous variables, as well as an ever-increasing computational power, modern hydrology is still facing major challenges. In particular, the interacting physical, chemical and biological aspects of the water cycle, their high spatial and temporal variability, as well as the fast growing role of anthropogenic activities, render the study and modelling of the hydrological functioning of river basins increasingly difficult (Ambroise, 1999). Having acknowledged these facts, Leonardo's studies on, and findings about the water cycle appear even more outstanding.

As shown in Chapter 3, Leonardo had a good perception of atmospheric processes that drive the circulation of water. Evaporation, condensation and the formation of clouds, their migration under the effect of winds, as well as the formation of raindrops that ultimately fall towards the earth, are all processes that Leonardo understood well. Although he realised that solar heat is the driving force behind these processes, he unfortunately did not fully recognize their link with the surface and subsurface components of the water cycle. According to Leonardo, the latter had their own driving forces and were only partially connected to the atmospheric water cycle.

Although he always insisted on the disconnection of the atmospheric and terrestrial parts of the water cycle, Leonardo revised his ideas of the terrestrial part several times during his life.

4.2 THE EARTH AS A LIVING BODY

As early as 1610, Francis Bacon was surprised to see the similar shapes of the African and South American continents. In 1658, the Reverend Père Placet published his work entitled: *Where it is shown that before the deluge, America was not separated from the other parts of the world* (Derruau, 1988). However, it was only more recently, in the second half of the 19th century by Snider, and at the beginning of the 20th century by Taylor (1910) and Baker (1911), that the foundations of the theory of continental drift were established (Derruau, 1988).

Wegener relied on these works for the elaboration of his famous concept on the continental drift published in 1915 in his book *The Origin of the Continents and of the Oceans*.

Relying strongly on Wegener's work and making use of the progress achieved by numerous studies throughout the first half of the 20th century, several researchers (McKenzie & Parker, 1967; Le Pichon, 1968; Morgan, 1968) almost simultaneously proposed the concept of plate tectonics, explaining the earth's structural components and the mechanisms by which they evolve. Large-scale thermal convection currents move the plates of the upper mantle. As they are either composed of continental or oceanic crust, these plates are bound by margins that are of three types (Whittow, 2000): (i) constructive, where on each side of the mid-oceanic ridges new ocean floor is created; (ii) destructive, where there is loss of ocean floor below a subduction zone; and (iii) conservative, when plates pass each other in a lateral movement, without causing any creation or destruction of ocean floor.

The lithosphere, i.e. the earth's crust, is subject to processes referred to as tectonogenesis, which cause deformations contributing to orogenesis. The latter includes folding, faulting and thrusting, resulting from plate tectonics, which buckle and deform sediments within geosynclines, compressing them into long linear mountain chains.

Leonardo tried on many occasions to develop a conceptual framework for the structure of what is considered today as the lithosphere. One of the images that was closest to Leonardo's heart is the Platonic view on the correspondence between human microcosms and macrocosms of nature. He very often comes back to this point comparing the earth to a living body, with its blood, flesh and bones: "*we might say that the earth has a spirit of growth; that its flesh is the soil, its bones the arrangement and connection of the rocks of which the mountains are composed, its cartilage the tufa* (soft rock), *and its blood the veins of water. The pool of blood which lies round the heart is the ocean, and its breathing, and the increase and decrease of the blood in the pulses, is represented in the earth by the ebb and flow of the sea; and the heat of the spirit of the world is the fire which pervades the earth, and the seat of the vegetative soul is in the fires, which in many parts of the earth find vent in baths and mines of sulphur, and in volcanoes, as in Mount Etna on Sicily, and in many other places*" (Codex Leicester).

In the above figurative description, Leonardo refers to the heat of the spirit of the world, meaning the source of energy that drives the water cycle, which in the end enabled the formation of life on our planet. He identified the volcanoes as manifestations of this source of energy. As we shall see later, in Chapter 5, Leonardo knew about solar energy driving the evaporation of water from the surface of the earth towards the atmosphere, but as stated in Chapter 2, he envisaged another source of energy driving the water inside and on the surface of the earth, the so-called "*element of fire*", located beyond the atmosphere.

Leonardo (unfortunately) adopted the ideas of Aristotle who believed that the world was composed of four elements arranged in concentric spheres, with the earth at its centre, progressively surrounded by water, air and fire: "*the ancients define man as the little world, and this definition is appropriate, since man is composed of earth, water, air and fire which is similar to the world. As man has bones, which sustain the flesh, the world has stones, which sustain the earth. As man has in him a pool of blood in which the lungs rise and fall in breathing, so the body of the earth has its ocean tides, which likewise rise and fall every six hours, as if the world breathed. As the pool of blood gives rise to veins, which branch out through the human body, similarly the ocean fills the earth with infinite veins of water. The earth's body does not have nerves, which are not there because they are made for movement, and since the earth is stable, it does not move, and since it does not move it does not need nerves. But in all other things they are very similar*" (Codex Paris, A).

By assuming the earth to be a living body, Leonardo set the contours for his study of the physical structure of our planet. Based on this concept, he would develop further theories on the constant reshaping of the surface of the earth, which he subsequently used as the framework for his water studies.

4.3 THE STRUCTURE OF THE EARTH

Concerning the structure of the earth, Leonardo recognised well that our planet is a sphere, but does not have a perfectly round shape: *"the centre of the sphere of waters is the true centre of the globe of our world, which is composed of water and earth, having the shape of a sphere. If you want to find the centre of the earth, this is placed at a point equidistant from the surface of the ocean, and not equidistant from the surface of the earth; for it is evident that this globe of earth has nowhere a perfect roundness, except in places where the sea is, or marshes or other still waters. And every part of the earth that rises above the water is farther from the centre"* (Codex Paris, A).

It is clear that Leonardo understood that if a mass could move freely, it would assume a position of equal potential in relation to the centre of the earth: a sphere. But he also identified the shaping force of erosion, which is needed to form a sphere: *"Water erodes mountains and fills valleys and would like to reduce the earth to a perfect sphere, if it could"* (Codex Atlanticus).

In a further explanation, Leonardo wrote: *"Let the earth turn on which side it may, the surface of the waters will never move from its spherical form, but will always remain equidistant from the centre of the globe. Even if the solid earth would be removed from the centre of the globe, what would happen to the water? It would remain in a sphere round that centre equally thick* (i.e. with the same volume), *but the sphere would have a smaller diameter than when it enclosed the* (solid) *earth"* (Codex Leicester).

Leonardo greatly underestimated the proportion of the earth submerged by ocean waters, as is demonstrated by his statement saying that *"some assert that it is true that the earth, which is not covered by water is much less than that covered by water. But considering the size of 7000 miles in diameter which is that of this earth, we may conclude the water to be of small depth"* (Codex Leicester).

Aristotle claimed that since the earth is heavier than water, it should be contained in it, and therefore be lower than it. However, it was evident that earth stands above sea level and therefore above the sphere of water. While the experts of his times tried to find a justification for this contradiction, thinking that the emerged earth was still below the sphere of water, Leonardo believed in what he saw and wrote: *"many people have arrogantly stated that the surface of the sea is higher than the highest mountain that exists"* (Codex Leicester).

He wrote about *"the relative height of the surface of the sea to that of the land"*, explaining that if *"b–d (see Fig. 4.1) is a plain through which a river flows to the sea; this plain ends at the sea, and since in fact the dry land that is uncovered is not perfectly level, for if it were the river would have no motion, as the river does move, this place is a slope rather than a plain; hence this plain b–d so ends where the sphere of water begins that if it were extended in a continuous line to b–a it would go down beneath the sea, whence it follows that the sea a–c–b looks higher than the dry land"*. He concluded that *"obviously no portions of dry land left uncovered by water can ever be lower than the surface of the water sphere*" (Fig. 4.1).

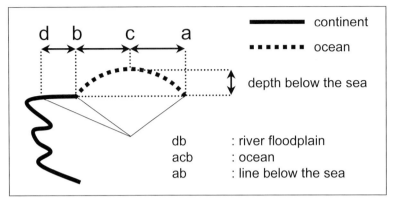

Fig. 4.1 *Diagram based on an illustration for a note on "the relative height of the surface of the sea to that of the land".*

As is apparent from these writings, Leonardo's view of the spherical form of our planet, with only the surface of the oceans at equal distance from the earth's centre, while the land emerging above those waters remains, of course, at further distance from the centre, was correct. Leonardo attempted to go further in his conceptualisation of the structure of the earth and especially the circulation of water within it. In this respect, he visualized a network of veins present inside the earth. Forming endless branching networks, those veins would be filled with water, moving from one part of the planet to another. Leonardo thus stated that "*the body of the earth, like the bodies of animals, is intersected with branching veins of waters which are all in connection and are constituted to give nutriment and life to the earth and to its creatures*" (Codex Leicester). In this conceptualisation, he considered this network of veins filled with flowing water as the place where the major part of the water cycle takes place, saying that "*these (the waters) come from the depth of the sea and, after many revolutions, have to return to it by the rivers created by the bursting of these springs*" (Codex Leicester). According to Leonardo, rivers stem from the veins that are close to the earth's surface and which are progressively eroded by the moving waters, like blood seeping from a wound. This erosion process was supposed to culminate in the crumbling of a vein and the outpouring of water from the damaged vein. Leonardo was convinced that these outbursts of water were the main cause of springs, giving birth to rivers.

It is interesting to note that Leonardo clearly understood the role of erosion as a force that is constantly remodelling the surface of the earth. In a way, he came close to the theory of erosion cycles formulated by W. M. Davis by the end of the 19th century. Davis's concept, however, had a major flaw, in that it supposed that a full cycle included an initial, long phase of erosion causing an almost perfect flattening of the relief, followed by a second very short phase of orogenesis. Leonardo's concept was clearly closer to reality in that it allowed for erosion and orogenesis to take place simultaneously. As we will see hereafter, he nevertheless was wrong in his concept of the forces that he believed were driving these processes.

4.4 WATER AS A FORCE ERODING AND SHAPING THE EARTH'S SURFACE

In his earliest conceptualisation of the earth's structure and the circulation of water, Leonardo identified the eroding power of running water, acting directly on the shape of the planet's surface. In order to explain why mountains rise above the sphere of water, so contradicting the evidence that the eroding power of water would "*reduce the earth to a perfect sphere*" (Codex Atlanticus), he imagined that initially the earth was surrounded by water, and that at the centre of the world there was a big cavern, full of water. Then he considered that "*a vast part of the vault of the cavern collapsed towards the centre of the earth, broken by the veins, which continuously erode the place where they flow*" (Codex Leicester). This would imply a redistribution of weight that caused the mountains to rise above the sphere of water (Fig. 4.2).

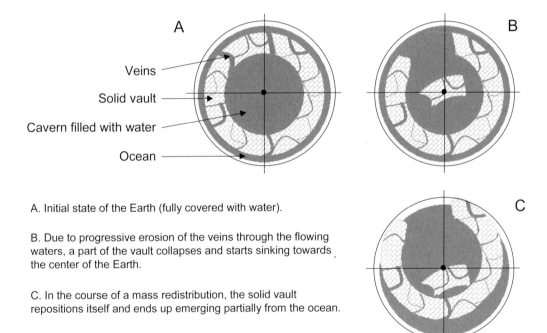

A. Initial state of the Earth (fully covered with water).

B. Due to progressive erosion of the veins through the flowing waters, a part of the vault collapses and starts sinking towards the center of the Earth.

C. In the course of a mass redistribution, the solid vault repositions itself and ends up emerging partially from the ocean.

Fig. 4.2 Illustration of Leonardo's concept for the formation of mountains emerging from the water sphere.

In Section 4.7 it will be shown that Leonardo did not consider the erosion of the subterranean veins and the related collapse of portions of the solid vault as the only force that is at the origin of the remodelling of the earth's surface. He believed that surface erosion indeed also played a significant role through the massive weight redistribution that it causes.

Although this concept of mountain formation is totally wrong, in the light of current knowledge on tectonics, Leonardo did have a good perception of the geomorphological processes that continuously rebuild the earth's surface and of the large amounts of groundwater present under the surface. He wrote that *"mountains are made by the currents of rivers. Mountains are destroyed by the currents of rivers"*. He recognised and described the importance of processes like erosion and sedimentation as demonstrated by his statement *"that the rivers have all cut and divided the mountains of the great Alps one from the other. This is visible in the order of the stratified rocks, because from the summits of the banks, down to the river the correspondence of the strata in the rocks is visible on either side of the river. That the stratified stones of the mountains are all layers of clay, deposited one above the other by the various floods of the rivers. That the different size of the strata is caused by the difference in the floods – that is to say greater or lesser floods"* (Codex Leicester).

In these writings we recognise fundamental theories of geomorphology. He clearly refers to geomorphological processes and their influence on the landforms upon which they are acting. The main process that Leonardo refers to is erosion, i.e. the wearing away of the land surface by the mechanical action of debris which is transported by the agents of erosion, including water and wind. In this text, he also recognises that mountains contain layers of sedimentary rock that were deposited when submerged by water; a knowledge which only reappeared in the 18th century. Leonardo did not refer to weathering, which is known today as a prerequisite for erosion. Weathering does not involve transport, but generates regolith that remains *in situ* until agents of erosion cause its transport.

Leonardo further identified erosion as a major creator of landforms, saying that *"the summits of mountains for a long time rise constantly. The opposite sides of the mountains always approach each other below; the depths of the valleys which are above the water sphere are in the course of time constantly getting nearer to the centre of the world"* (Codex Paris, L).

In a similar way, he also wrote that *"in an equal period, the valleys sink much more than the mountains rise"* (Codex Paris, L). Likewise, he stated that *"the bases of the mountains always come closer together"*, and that *"in proportion as the valleys become deeper, the more quickly are their sides worn away"* (Codex Paris, L). This last statement demonstrates that Leonardo had also identified the importance of slope to the efficiency and speed of erosion. The steeper a slope, the greater the force of gravity acting on the loose material, ultimately causing its descent into the valley.

Leonardo recognised well the link between erosion on mountain peaks and slopes and the deposition of the eroded material in the valleys: *"The base of the mountains may be in great part clothed and covered with ruins of shrubs, hurled down from the sides of their lofty peaks, which will be mixed with mud, roots, boughs of trees, with all sorts of leaves thrust in with the mud and earth and stones"* (Codex Leicester).

He further refers to the occasional reshaping of the valley bottoms by the flowing waters of rivers during floods: *"And into the depth of some valley may have fallen the fragments of a mountain forming a shore to the swollen waters of its river; which, having already burst its banks, will rush on in monstrous waves; and the greatest will strike upon and destroy the walls of the cities and farmhouses in the valley*" (Codex Leicester).

In the same context, Leonardo tried to explain the differences that can be observed between river courses. He argued that of two linear rivers flowing in a homogenous region, with the same abundance of water and equal width, length, depth and slope, the eldest one is also the slowest one. According to his theory, the most sinuous river course is the oldest one and it will progressively reduce its speed. This is a remarkable observation, which indicates that Leonardo had a good perception of the time scale of landscape formation. Nowadays we see this as maximization of entropy. Mature rivers have reached higher levels of entropy with more gradual and evenly spread energy dissipation.

Interestingly, Leonardo thought that the confluence of two rivers in a large plain is actually the cause of meandering (Fig. 4.3). This is why he advised those who wanted to build canals to make sure that they connected the smaller rivers to the larger ones at a sharp angle. This would ensure that the current of the larger river deflects the entry line of the smaller river, thus preventing the latter from "hitting" the opposite bank.

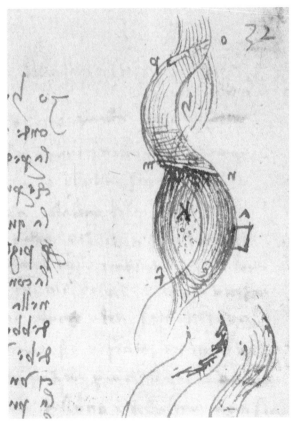

Fig. 4.3 *Detail showing currents in a meandering river (c. 1508–1512). Codex Leicester. © Seth Joel/CORBIS.*

Moreover he wrote that meanders of a river can be destroyed during major floods, due to the greater force of the current, forcing water to flow forward in a straight direction.

Leonardo presents his thoughts on meandering rivers in detail when he comments on Fig. 4.4, writing that "*When the fullness of rivers is diminished* (i.e. when the water level drops)*, then the acute angles formed at the junction of their branches become shorter at the sides and wider at the point; like the current a–n and the current d–n, which unite at n when the river is at its greatest fullness. I say, that when it is in this condition if, before the fullest time, d–n was lower than a–n, at the time of fullness d–n will be full of sand and mud. When the water d–n falls, it will carry away the mud and remain with a lower bottom, and the channel a–n finding itself the higher, will fling its waters into the lower, d–n, and will wash away all the point of the sand-spit b–n–c, and thus the angle a–c–d will remain larger than the angle a–n–d and the sides shorter, as I said before*" (Codex Leicester).

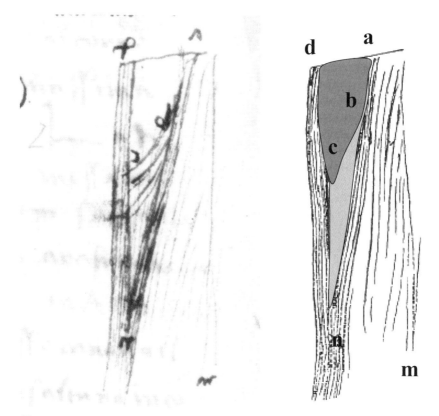

Fig. 4.4 Study of riverbed erosion and sedimentation (c. 1508–1512). Detail from sheet discussing river control techniques, Codex Leicester. © Seth Joel/CORBIS.

While observing the erosion and sedimentation processes in a riverbed, Leonardo tried to draw conclusions from them, allowing him to correctly explain the driving forces behind the perpetual remodelling of the river courses: "*the more rapid it* (the water) *is, the more it wears away its channel; and, by the converse proposition, the slower the water the more it deposits that which renders it turbid*" (Codex Arundel). This observation sounds logical, but is not entirely true. It is correct to say that when the water accelerates, it erodes, and when it slows down, it deposits sediment.

Reflecting on Leonardo's views on the mechanisms causing meandering of rivers, we need to realise that even today the exact cause of meandering is not yet perfectly understood. Meandering is believed to be the most probable state achieved (maximum entropy) when a river exercises the least amount of work per unit volume of water. Since a meander lengthens the river course, it reduces the gradient of a river, thereby reducing energy expenditure, despite increasing the volume of water being moved. Meanders tend to migrate downstream, with bank erosion occurring where the velocity is greatest (on the outside of the bends) and point-bar deposition occurring where the velocity is least (on the inside of the bends) (Whittow, 2000).

4.5 THE ROLE OF SEDIMENTATION IN THE BUILDING OF LANDFORMS

Besides identifying the potential of erosion to shape landscapes, Leonardo also developed a clear concept on sedimentation in the lower parts of the landscape. In his writings he described how the structure of the earth's crust was directly linked to the processes of erosion and sedimentation: "*Every part of the depth of earth in a given space is composed of layers, and each layer is composed of heavier or lighter materials, the lowest being the heaviest. And this can be proven, because these layers have been formed by the sediment from water carried down to the sea, by the current of rivers which flow into it. The heaviest part of this sediment was that which was first thrown down, and so on, by degrees; and this is the action of water when it becomes stagnant, having first brought down the mud whence it first flowed. And such layers of soil are seen in the banks of rivers, where their constant flow has cut through them and divided one slope from the other to a great depth; where in sedimentary layers the waters have drained out, the materials have, in consequence, dried and converted into hard stone, and this happened most in what was the finest mud; whence we conclude that every portion of the surface of the earth was once at the centre of the earth, and vice-versa*" (Codex Arundel).

The quote above shows that Leonardo had perceived the huge amount of information that can be found in the layers of alluvial deposits. Indeed, sedimentary deposits, as we go downstream, exhibit a marked decrease in particle size, mainly due to sorting (i.e. grading of material according to their particular attributes, such as shape, density, size). He wrote that "*a river that flows from a mountain, deposits a great quantity of large stones in its bed, which still have some of their angles and sides, and in the course of its flow it carries down smaller stones with the angles more worn; that is to say the large stones become smaller. And further on it deposits coarse gravel and then smaller, and as it proceeds this becomes coarse sand and then finer, and going on, the water, turbid with sand and gravel, joins the sea; and the sand settles on the sea-shores, being cast up by the salty waves; and there the sand has so fine a nature as to seem almost like water, and it will not stop on the shores of the sea but return by reason of its lightness*" (Codex Leicester).

He went much further in his description of the mechanical forces that act on eroded rock fragments as they are transported by flowing waters: "*All the torrents of water flowing from the mountains to the sea carry with them the stones from the hills to the sea, and by the influx of the seawater towards the mountains; these stones were thrown back towards the mountains, and as the waters rose and retired, the stones were tossed about by it and in rolling, their angles hit together; then as the parts which least resisted the blows, were worn off, the stones ceased to be angular and became round in form, as may be seen on the banks of the Elsa. And those remained larger which were less removed from their native spot; and they became smaller, the farther they were carried from that place, so that in the process they were converted into small pebbles and then into sand and at last into mud*" (Codex Leicester).

Although Leonardo correctly observed the sorting of sediments along a river, he misunderstood the underlying mechanism of sorting. He perceived that sorting is caused only by the continued erosion of large particles to smaller particles, whereas the sorting is primarily the result of the decreasing transport capacity of the stream as its slope gradually reduces downstream. Leonardo also perceived the link between sedimentation layers and the formation of solid rocks. Indeed, he wrote that: "*still, being almost of the nature of water itself, it afterwards, when the weather is calm, settles and becomes solid at the bottom of the sea, where by its fineness it becomes compact and by its smoothness resists the waves which glide over it; and in this shells are found; and this is white earth, fit for pottery*" (Codex Leicester).

Leonardo further described the formation of rocks: "*the brine left by the sea with other humours of the earth made a concretion of the pebbles and the sand, so that the pebbles were converted into rock and the sand into tufa*" (soft rock) (Codex Leicester). This description corresponds to the formation of what are referred to nowadays as sedimentary rocks (see Chapter 4 frontispiece), i.e. rocks formed by the accumulation of material derived from pre-existing rocks or from organic sources.

Probably the most important fact that led Leonardo to his evolutionary view of the formation of the earth is the presence of fossils in rock layers. At his time, there were many theories that attempted to explain the occurrence of fossils in the earth's crust. Well before Leonardo, Avicenna (AD 980–AD 1037) and Aristotle had suspected the marine origin of fossils, but others believed that they were transported by the biblical deluge, or that they could grow within the earth without participating in organic life, or that they had a connection with extraterrestrial influences, or that they were a result of unsuccessful attempts at creation.

For the first time in history, Leonardo understood that fossils are a result of deposition and stratification of sediment layers at the bottom of ancient seas, and he understood that, as a result of geological movements, the bottom of those seas had been moved to the tops of high mountains: "*the shells, oysters, and other similar animals, which originate in sea-mud, bear witness to the changes of the earth round the centre of our elements. This is proved thus: Great rivers always run turbid, being coloured by the earth, which is stirred by the friction of their waters at the bottom and on their shores; and this wearing disturbs the face of the strata made by the layers of shells, which lie on the surface of the marine mud, and which were produced there when the salt waters covered them; and these strata were covered over again from time to time, with mud of various thickness, or carried down to the sea by the rivers and floods of more or less extent; and thus these layers of mud became raised to such a height, that they came up from the bottom to the air. At the present time these bottoms are so high that they form hills or high mountains, and the rivers, which wear away the sides of these mountains, uncover the strata of these shells, and thus the softened side of the earth continually rises and the antipodes sink closer to the centre of the earth, and the ancient bottoms of the seas have become mountain ridges*" (Codex Paris, E).

Evidently, Leonardo recognised that the shells he found in the mountains actually had their origin in a marine environment that existed long before his time. He emphasised the importance of these shells in understanding the formation of the earth's surface in his description of: "*the authorities for the study of the structure of the earth. Since things are much more ancient than letters, it is no marvel if, in our day, no records exist of these seas having covered so many countries; and if, moreover, some records had existed, war and conflagrations, the deluge of waters, the changes of languages and of laws would have consumed everything ancient. But sufficient for us is the testimony of things created in the salt waters, and found again in high mountains far from the seas*" (Codex Leicester).

Leonardo mainly referred to deposits resulting from transport by rivers to the oceans. According to the various soil types and rocks that were eroded by these rivers, he identified different sedimentary layers with regard to structure, type and origin of the deposited material. Leonardo believed that these layers were progressively lifted to their current altitudes, and thus were progressively undergoing erosion. In the following section we discuss Leonardo's hypothesis to explain the rise of these layers to their current altitudes.

4.6 LEONARDO'S CONCEPT OF EROSION AND THE FORCES SHAPING THE EARTH'S SURFACE, AS OPPOSED TO THE BIBLICAL DELUGE THEORY

By stating that "*every portion of the surface of the earth was once at the centre of the earth, and vice-versa*", Leonardo shows that he had understood the never-ending cycle of relief lifting and erosion. He was particularly eager to understand and describe the formation of high mountain relief. While doing so, he systematically referred to the existence of fossils at high locations in the Alps. Leonardo opposed the theory that the fossils were an evidence of the biblical deluge, arguing that there was no objective proof that could justify this assumption. He wrote that one has "*first to prove that the shells at a thousand braccia* of elevation were not carried there by the deluge, because they are seen to be all at one level, and many mountains are seen to be above that level; and to inquire whether the deluge was caused by rain or by the swelling of the sea; and then you must show how, neither by rain nor by swelling of the rivers, nor by the overflow of this sea, could the shells – being heavy objects – have floated up the mountains by the sea, nor have been carried there by the rivers against the course of their waters*". He was basically searching for an explanation to

---

* A braccio (pl. braccia) is an old measurement unit, approximately equal to the distance from the shoulder to the wrist, which was used in Renaissance Italy. Its length usually ranged from 66 to 68 cm, but it is also reported for having varied between 46 and 71 cm.

the fundamental question of "*Why do we find the bones of great fishes and oysters and corals and various other shells and sea-snails on the high summits of mountains, just as we find them in low seas?*" (Codex Leicester).

It is clear that Leonardo recognised that fossils, or shells as he called them, could not have been transported to high altitudes by means of what he referred to as the biblical deluge. He did not directly negate the existence of the deluge, which would probably have been a delicate political issue at his time, but he tried instead to give countless examples and arguments to prove that it simply could not have taken place.

One of his arguments questioned how the waters that had covered the entire earth could have disappeared after the deluge, since no downward flow was possible any more (as no relief emerged from the surface of the oceans). The only possibility that remained was that "*all this water was evaporated by the heat of the sun*".

In another of his writings on the biblical deluge, Leonardo argued that the fact that many marine animals only move at very low speed proves that they could never have "*travelled from the Adriatic sea as far as Monferrato in Lombardy, which is 250 miles distance, in 40 days* (i.e. the duration of the biblical deluge)" (Codex Leicester). For Leonardo it was obvious that the locations where the shells were found, in the mountains, must have been the same places as where these marine animals actually lived, i.e. the bottom of the sea close to the shores. A major proof for this assumption was the fact that in those locations one can find: "*sea-snails, and cuttlefish, and all the other shells which congregate together, all together and dead; and the solitary shells are found wide apart from each other, as we may see them on sea-shores every day*" (Codex Leicester).

Furthermore, he argued that the shells could never have been carried far by the waves of the deluge, saying "*where the dead went they were not far removed from the living* (Codex Leicester). Leonardo was convinced that the shells in the mountainous areas could never have been brought there by the deluge, since "*the deluge was caused by rain water, so that all these waters ran to the sea, and the sea did not run up the mountains; and as they ran to the sea, they thrust the shells from the shore of the sea and did not draw them towards themselves*" (Codex Leicester).

Besides arguing against the deluge as the reason for the occurrence of fossil shells in mountainous areas, Leonardo also opposed theories implying that fossils were a creation of God, writing about "*ignorant persons declaring that Nature or Heaven created the shells in these places by celestial influences, as if in these places we did not also find the bones of fishes which have taken a long time to grow; and as if we could not count in the shells of cockles and snails the years and months of their life, as we do in the horns of bulls and oxen, and in the branches of plants that are cut*" (Codex Leicester).

His reasoning is beautifully summarised in one rhetoric question: "*Why do we find so many fragments and whole shells between layer and layer of stone, if this had not formerly been covered on the shore by a layer of earth thrown up by the sea, and which was afterwards petrified? ... And between these separate layers of rock they are found, few in number and in pairs like those which were left by the sea, buried alive in the mud, which subsequently dried up and, in time, were petrified*" (Codex Leicester).

Here, Leonardo also describes the process of sedimentation and fossilisation that created the sedimentary rocks in which the fossils are found. In order to explain how those sedimentary rocks had been lifted to their current altitudes in mountainous areas, Leonardo stated that: "*if the earth of our hemisphere is indeed raised by a certain amount higher than it used to be, it must have become lighter by the same amount through the waters which it lost through the rift between Gibraltar and Ceuta; and all the more the higher it rose, because the weight of the waters which were thus lost would be added to the earth in the other hemisphere. And if the shells had been carried by the muddy deluge, they would have been mixed up, and separated from each other amidst the mud, and not in regular steps and layers, as we see them now in our time*" (Codex Leicester). As shown in Section 4.4, Leonardo believed that a complex redistribution of weight through movements of water and rocks was at the origin of what is nowadays considered as tectonics, i.e. internal forces which deform the earth's crust and thus create landforms (or tectonic relief), which are created by warping or fracturing.

4.7 EROSION AS A CONSTANT REMODELLING FORCE OF THE EARTH'S SURFACE UNDER THE EFFECT OF TECTONICS

Leonardo identified erosion as a never-ending cycle, transforming high mountains into flat plains and *vice versa*. He wrote of the beginning of the erosion cycle: "*that of old* (i.e. very long times ago)*, the state of the earth was that its plains were all covered up and hidden by salt water*" (Codex Atlanticus). Leonardo described a part of this erosion cycle for the Mediterranean region: "*the shores of the sea constantly bring more soil towards the middle of the sea; that the rocks and outcrops in the sea are constantly being ruined and worn away; that the Mediterranean sea will in time expose its bottom to the air, and all that will be left will be the channel of the greatest river that enters it; and this will run to the ocean and pour its water into it with those of all the rivers that are its tributaries*" (Codex Leicester).

Leonardo had a very complex theory which involved the continuous redistribution of weights over the entire globe. He rated the surface erosion of soils and rocks as being part of the major natural forces that characterise his redistribution theory, stating: "*the persistent wearing of the rivers cut through the mountains and the wandering courses of the rivers carried away the plains enclosed by the mountains; and the cutting away of the mountains is evident from the strata in the rocks, which correspond in their sections as made by the courses of the rivers*" (Codex Leicester).

Leonardo here describes a concept that was first proposed as a scientific theory by W. M. Davis in 1889. Davis's concept deals with the cyclic evolution and modification of the physical landscape, where the various stages of erosion are believed to be parts of an orderly cycle. Nowadays, this concept of cyclic progression is not well regarded, as far as landscape development is concerned, partly because complete levelling by rivers is probably never achieved and partly because varying climatic conditions imply that processes do not remain constant in a given region.

From Leonardo's perspective, those places on earth where the river network has the highest density are subject to the largest loss of sediment, which causes a loss in weight, resulting ultimately in a rise of the relief and the formation of mountains. He stated that "*great rivers always run turbid, being coloured by the earth, which is stirred by the friction of their waters at the bottom and on their shores*" (Codex Paris, E). In one sense, Leonardo here indirectly refers to the principle of tracer hydrology, whereby flowing water always reflects the physical and chemical signature of the drainage basin, through the sediments and chemical elements it receives from flowing through soils and rocks. In Section 5.3.5 we detail how Leonardo understood the concept of tracers, and how he described this in poetic terms.

According to Leonardo, flowing water, mainly through its eroding effect, causes the earth to become progressively lighter. Thus, *"that part of the earth which was lightest remained farthest from the centre of the world; and that part of the earth became the lightest over which the greatest quantity of water flowed. And therefore that part became lightest where the greatest number of rivers flow; like the Alps which divide Germany and France from Italy; from where the Rhone flows Southwards, and the Rhine to the North. The Danube or Tanoia towards the North East, and the Po to the East, with innumerable rivers which join them, and which always run turbid with the soil carried by them to the sea"* (Codex Leicester).

Furthermore, he noted that the eroded sediments are carried to the Mediterranean Sea by the rivers, causing a progressive filling of the marine sink: *"the shores of the sea are constantly moving towards the middle of the sea and displace it from its original position. The lowest portion of the Mediterranean will be reserved for the bed and current of the Nile, the largest river that flows into that sea"* (Codex Leicester).

Leonardo's ideas of continuous erosion and sedimentation, and his view on the constant reshaping of the earth's crust, were well ahead of more recent concepts such as those pronounced by W. M. Davis, a little more than a century ago. The movement of water clearly was part and parcel of his morphological theories. As we will see in the following chapters, Leonardo dedicated a tremendous amount of work to the study of water, either as part of the global water cycle or as water flowing through a river or stream, causing permanent changes to its cross-sections.

5 THE WATER CYCLE

Deluge over an alpine valley (c. 1506).
The Royal Collection © 2008 Her Majesty Queen Elizabeth II (RL 12409r).

5.1 THE GENERAL FRAMEWORK FOR LEONARDO'S CONCEPT OF THE WATER CYCLE

The modern concept of the water cycle describes the movement of water through the earth–atmosphere system, driven by solar energy which evaporates water from water and land surfaces (including vegetation), transports it over large distances, and releases it by condensation (clouds) from where it precipitates on land and water surfaces in the form of rain, snow, hail or dew. At the earth's surface, the precipitation is stored on the surface (rivers, lakes, glaciers) or below ground as soil moisture and groundwater, or it is evaporated to feed the "next" cycle, with the remainder being returned to the sea by river flow and groundwater seepage.

Leonardo dedicated much work to the water cycle, trying to figure out approaches and concepts for explaining the origin of flowing water, lakes and oceans on the earth. As we have seen in the previous chapters, he had clearly identified the link between erosion cycles and the water cycle, referring to the upward movement of land under the effect of a loss of weight due to the progressive erosion of soils and rocks by the waters flowing on the earth's surface, as well as inside the solid crust. One of his most pertinent statements refers to the fact that "*whenever discussing water, you first have to invoke experience, before referring to reason*" (Codex Paris, H). His appreciation of the importance of field observation and experiment aimed at elucidating the processes before setting up theories, shows how close he came to being a true scientist in the modern sense of the word.

Leonardo also referred to the study of the motion of water, as well as the environment in which this movement takes place, saying that one had to "*first write of all water, in each of its motions; then describe all its bottoms and their various materials, always referring to the propositions concerning the said waters; and let the order be good, for otherwise the work will be confused ... Describe all the forms taken by water from its greatest to its smallest wave, and their causes*" (Codex Paris, F). Once again, Leonardo tries to set the overall context of the power of erosion of water, supposing that this was the initial cause of the formation of landforms.

Leonardo's conceptual framework of the water cycle was further developed in his treatise on water, through the use of a figurative example. As already seen above, he adhered to a theory, which he called a concept of the ancients: "*comparing man to the world in miniature*". Leonardo fully believed in this concept, saying that "*inasmuch as man is composed of earth, water, air and fire, his body resembles that of the earth*" (Codex Paris, A).

He gave a number of examples to further illustrate the similarity between the earth and the human body, saying that (Codex Paris, A):

- "*As man has in him bones the supports and framework of his flesh, the world has its rocks the supports of the earth*".
- "*As man has in him a pool of blood in which the lungs rise and fall in breathing, so the body of the earth has its ocean tide which likewise rises and falls every six hours, as if the world breathed*". This original concept of the origin of ocean tides is discussed in Section 5.3.6.
- "*As in that pool of blood veins have their origin, which ramify all over the human body, so likewise the ocean sea fills the body of the earth with infinite springs of water*". This is a fundamental part of Leonardo's concept of the water

cycle, establishing a direct link between the oceans and the rivers that run on the earth's surface, interconnected through a network of veins that exist throughout the terrestrial crust.

The only major difference between the earth and the human body in this comparison is, according to Leonardo, the absence of sinews and muscles within the earth, "*because the sinews are made exclusively for movement and, the world being perpetually stable, no movement takes place, and no movement taking place, muscles are not necessary*" (Codex Paris, A). Leonardo insisted on his conviction that "*in all other points they* (the earth and the human body) *are much alike*" (Codex Paris, A). Being such a capable observer in all the fields of physics and engineering in which he excelled, it may seem strange to us now that Leonardo embraced this erroneous concept, particularly since there had been earlier classical authors who had rejected this theory (e.g. Anaxagoras, 5th century BC). But we should not forget that Leonardo was also an artist, to whom the poetic concept of a living world, so close to the present-day Gaia concept (Lovelock, 1979), must have seemed very appealing as an all-encompassing theory.

5.2 LEONARDO'S KEY QUESTIONS RELATING TO THE WATER CYCLE

Having developed this detailed conceptual framework of the water cycle, based on the similarity between the earth and the human body, Leonardo defined a series of main issues in the study of the water cycle, which were intended to be part of his "*Treatise on Water*" (Codex Paris, E):

- "*Define first what is meant by height and depth; also how the elements are situated one inside another*". He considered height differences a necessary condition for the flow of water towards lakes and oceans, and thus also for the erosion of soils and rocks.
- "*Then, what is meant by solid weight and by liquid weight; but first what weight and lightness are in themselves*". This statement refers to his concept of vertical movements of the earth's crust, driven by weight differences. Under the effect of erosion, the loss of weight via the transport of eroded material from the mountains to the oceans would inevitably cause the lighter parts of the earth to move upwards.
- "*Then describe why water moves, and why its motion ceases; then why it becomes slower or more rapid; besides this, how it always falls, being in contact with the air but lower than the air*". Leonardo recognised that there needs to be an engine that drives the water cycle. Many of his writings on the water cycle deal with the issue of identifying and describing those driving forces.
- "*And how water rises in the air by means of the heat of the sun, and then falls again in rain*". He was close to correctly understanding the main engine driving the water cycle, i.e. the heat of the sun, but as shown in Section 2.3, he believed in an additional even more efficient source of energy, located beyond the outer limits of the atmosphere and which he called the "*element of fire*".
- "*Again, why water springs forth from the tops of mountains; and whether the water of any spring higher than the ocean can pour forth water higher than the surface of that ocean*". The transport of water towards the top of the mountains

appears as one of the major points of interest in Leonardo's quest for a correct understanding of the water cycle. As will be shown, the interconnection between the waters of the oceans and the rivers, via a subterranean network of veins, is one of his major explanations of this.

He also observed that in the seas there is no important circulation of water: "*sea is the name given to that water which is wide and deep, in which the waters have not much motion*".

Ultimately, it appears that most of Leonardo's questions relating to the water cycle can be summarised in the three key questions, formulated by Penman (1963) and Hewlett & Hibbert (1967), which still remain of great interest to contemporary hydrologists:

- *Where does the water go when it rains?*
- *What flow path does it take to the stream?*
- *How long does it reside in a catchment?*

While hydrology remained descriptive and empirical for a long time, the study of water pathways in catchments has seen tremendous progress over the past 30 years. New scientific approaches, both dynamic and systemic, have yielded new analytical tools and measuring devices that help to better understand the complexity of the water cycle (Ambroise, 1999). But still much more research is needed before we can safely say that we fully understand what happens to the rain as it travels through our catchments.

In the following, the ideas and concepts that Leonardo developed to find answers to the above questions will be explained in more depth.

5.3 LEONARDO'S CONCEPT OF THE WATER CYCLE

5.3.1 The idea of drainage basins

The concept of the drainage basin is one of the foundations of modern hydrology. It represents an open system, whose functioning in response to atmospheric forcing results in a temporal and spatial redistribution of incoming precipitation, tending to concentrate water along given pathways. For the study of the water cycle and the cycles coupled to water (energy, solutes, sediments, biomass, etc.), the drainage basin is a fundamental unit for the development, management and protection of water resources (Ambroise, 1999).

Leonardo gave numerous descriptions of what we would nowadays call drainage basins. He recognised the functioning of river networks, draining large areas, as illustrated by his description of the River Nile: "*the Nile comes from Southern regions and traverses various provinces, running towards the North for a distance of 3000 miles and flowing into the Mediterranean Sea by the shores of Egypt; and if we will give to this a fall of ten braccia a mile, as is usually allowed to the course of rivers in general, we shall find that the Nile must have its mouth ten miles lower than its source*" (Codex Leicester). In a similar way, he described the major river basins, draining the European continent: "*one can see the Rhine, the Rhone and the Danube starting from the German parts, almost at the centre of Europe, and having a course one to the East, the other to the North, and the last to southern seas*" (Codex Leicester).

The gravity-driven downstream flow of water was a concept that was far from being universally accepted during Leonardo's lifetime. In a commentary on this issue he wrote: "*of the sea, which to many fools appears to be higher than the earth which forms its shore*" (Codex Leicester).

5.3.2 The hydrological cycle

The document that is nowadays accepted as the first paper describing the hydrological cycle was published by Perrault in 1694, who showed that the measured rainfall was sufficient to sustain discharge of the upper Seine (Dooge, 1959).

The modern concept of the hydrological cycle considers the complex and cyclic interrelationships between all atmospheric and terrestrial processes related to moisture transport, such as evaporation, advection, condensation, precipitation, overland flow, infiltration, sub-surface flow, open channel flow and groundwater flow.

When writing about the hydrological cycle, Leonardo tried to give examples of what might happen if the water was not part of a never-ending cycle. From his viewpoint, the earth would soon dry out and ultimately there would be no life remaining: "*The rivers will be deprived of their waters, the fruitful earth will put forth no more her light verdure; the fields will no more be decked with waving corn; all the animals, finding no fresh grass for pasture, will die and food will then be lacking to the lions and wolves and other beasts of prey, and to men who after many efforts will be compelled to abandon their life, and the human race will die out*" (Codex Arundel).

Leonardo came close to the modern definition of the hydrological cycle when he referred to the water that passes through the major river systems countless times, adding up to volumes that are much greater than those contained in the oceans. He provided the example of the rivers Tigris and Euphrates, where the water must "*have flowed from the summits of the mountains of Armenia*" and where "*it must be believed that all the water of the ocean has passed many times through these mouths*" (Codex Paris, A). Leonardo further illustrated his concept of a never-ending hydrological cycle by the River Nile that "*must have sent more water into the sea than at present exists of all the element of water*" (Codex Paris, A).

In what can be considered as a very accurate description of the hydrological cycle, Leonardo wrote that: "*we may conclude that the water goes from the rivers to the sea, and from the sea to the rivers, thus constantly circulating and returning, and that all the sea and the rivers have passed through the mouth of the Nile an infinite number of times*" (Codex Paris, A).

From the above it can be concluded that Leonardo's perception of the hydrological cycle was correct in the sense that he recognised both the never-ending character of this cycle, and the role of river networks as drainage systems conducting the flow of water from the continents to the oceans.

Since he lacked the necessary knowledge of the components of the cycle (evaporation, condensation, precipitation and runoff), for a correct understanding of the entire hydrological cycle, Leonardo did not figure out their correct sequence and interrelationships. The major shortcoming in his concept lies in the perceived transfer of ocean water to the summit of the mountains, where it would ultimately start its downward movement all over again, under the effect of gravity.

Leonardo thought of a subterranean part of the hydrological cycle, consisting of a network of veins where the water would travel from the oceans to the summit of the mountains. He indeed wrote about *"very large rivers"* that *"flow under ground"* (Codex Atlanticus). In his concept *"the waters start from the bottom of the seas, and ramifying through the earth they rise to the summits of the mountains, flowing back by the rivers and returning to the sea"*.

In trying to explain the rise of waters from the oceans to the summits of the mountains, Leonardo referred to similarities between the earth and the human body on many occasions. In one of his notebooks he wrote that: *"the waters circulate with constant motion from the utmost depths of the sea to the highest summits of the mountains, not obeying the nature of heavy matter; and in this case it acts as does the blood of animals which is always moving from the sea of the heart and flows to the top of their heads; and here it is that veins burst, as one may see when a vein bursts in the nose, that all the blood from below rises to the level of the burst vein. When the water rushes out of a burst vein in the earth it obeys the nature of other things heavier than the air, whence it always seeks the lowest places. These waters traverse the body of the earth with infinite ramifications"* (Codex Leicester; see also Fig. 5.1).

Fig. 5.1 Exploding mountain: "water rushing out of a burst vein" (c. 1517–1518). The Royal Collection © 2008 Her Majesty Queen Elizabeth II (RL 12380r).

It is obvious that Leonardo considered it as his obligation to find out what was the driving force for the upward movement of waters. In this regard, he wrote that: "*the same cause which stirs the humours in every species of animal body and by which every injury is repaired, also moves the waters from the utmost depth of the sea to the greatest heights*" (Codex Arundel). He continues: "*it is the property of water that it constitutes the vital humour of this arid earth; and the cause which moves it through its ramified veins, against the natural course of heavy matters, is the same property which moves the humours in every species of animal body. But that which crowns our wonder in contemplating it, is that it rises from the utmost depths of the sea to the highest tops of the mountains, and flowing from the opened veins returns to the low seas; then once more, and with extreme swiftness, it mounts again and returns by the same descent, thus rising from the inside to the outside, and going round from the lowest to the highest, from whence it rushes down in a natural course. Thus by these two movements combined in a constant circulation, it travels through the veins of the earth*" (Codex Arundel).

Leonardo's statements on the hydrological cycle presented above have a major shortcoming, in that they clearly underestimate the enormous spatial and temporal variability of the hydrological cycle. His theories reflect his wish to find the key to a sort of universal understanding of the processes involved.

Modern hydrology has faced similar problems, in that certain mainstream theories on streamflow generation prevailed for decades. In his paper on the role of infiltration in the hydrological cycle, Horton (1933) identified infiltration capacity as a catchment average value, being constant in both space and time, and determining storm runoff as a fraction of rain falling at intensities exceeding the soil's infiltration capacity.

Although Horton later considered variable infiltration capacities over a catchment, it took many years before studies performed in numerous experimental catchments shed light on the complexity of runoff generation. This is illustrated by Betson's work (1964) questioning the concept of Hortonian surface runoff and insisting on the importance, but also difficulties, of "*measuring and understanding the physical factors controlling the infiltration process*". In 1970, Dunne & Black introduced the partial area contribution concept to explain storm runoff, based on saturation excess overland flow, but restricted to certain areas of the watershed. The variable source area concept was the dominating concept for decades. The spatial and temporal variability of the origin of the water encountered in stream discharge only became the subject of study some 30 years ago through the use of chemical and isotope tracers (see e.g. Pinder & Jones, 1969; Sklash & Farvolden, 1979).

As noted by McDonnell (2003), it is only recently that new insights into water sources, flow paths and residence times have provided the momentum for a major paradigm shift in hydrology, recognising that so-called pre-event water largely dominates storm runoff. Even though recent progress in experimental hydrology has led to new insights as to where the water goes when it rains, so-called universal concepts are still widely used. Still too often, hydrological models are dominated by a one-model-fits-all view, based on the outdated runoff generating mechanisms that were elaborated several decades ago (McDonnell, 2003).

How can we expect Leonardo, who lived more than 500 years ago at the beginning of the Renaissance, when no experimental catchments nor sophisticated computer systems and software existed, to have coped with these fundamental problems on how the hydrological system works and on how water is distributed in space and time. Keeping this in mind, we now investigate the solutions he proposed to solve the main questions of hydrology.

5.3.3 Leonardo's thoughts on "where does the water go when it rains?" and "what flow path does it take to the stream?"

While Leonardo recognised the never-ending character of the hydrological cycle, consisting basically in the rise of water to the top of the mountains to ultimately flow back to the seas, his concept of the water fluxes inside the hydrological cycle changed continuously. From the beginning, he perceived gravity as the driving force for the flow of waters down the slopes of the mountains, but he revised his explanation for the uplift of waters to the top of the mountains several times.

Leonardo interrogated himself on the fact that "*the water of the ocean cannot make its way from the base to the top of the mountains which bound it, but only so much rises as the dryness of the mountain attracts*" (Codex Paris, Manuscript G). Here we find a first attempt to explain the rise of water from the seas, which was supposed to take place through a sort of capillary rise from the base of the mountains (i.e. the level of the ocean). He also developed a theory based on the principle of a siphon, lifting the water from the base of the mountain to its top (refer to Fig. 5.2): "*And if, on the contrary, the rain, which penetrates from the summit of the mountain to the base, which is the boundary of the sea, descends and softens the slope opposite to the said mountain and constantly draws the water, like a siphon which pours through its longest side, it must be this which draws up the water of the sea; thus if s–n were the surface of the sea, and the rain descends from the top of the mountain a to n on one side, and on the other sides it descends from a to m, without a doubt this would occur after the manner of distilling through felt, or as happens through the tubes called siphons. And at all times the water which has softened the mountain, by the great rain which runs down the two opposite sides, would constantly attract the*

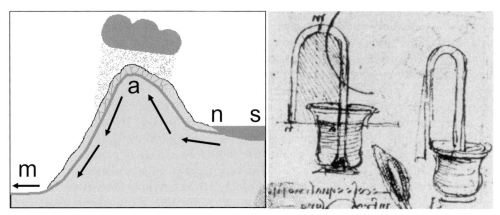

Fig. 5.2 Diagram relating to Leonardo's notes on the circulation of water between the oceans and the tops of mountains (left), based on the concept of a siphon (right). Detail from the Drawing of Nude Figures and Siphons, Codex Paris, Manuscript G. © Alinari Archives/CORBIS.

rain a–n, on its longest side together with the water from the sea, if that side of the mountain a–m were longer than the other a–n" (Codex Paris, G).

Leonardo's siphon concept is based on the branching network of veins that he supposed existed inside the earth. He was convinced that these networks formed a direct link between the oceans and the top of the mountains. Through the siphoning, he tried to establish this link and to find the driving force behind the uplift of water from the ocean to the mountains. However, after completing the description of the siphon concept, he appears to have been dissatisfied with this idea, writing that *"this cannot be, because no part of the earth which is not submerged by the ocean can be lower than that ocean"*.

While still holding on to his idea of water circulating inside the earth via a network of subterranean veins, Leonardo tried to find another explanation for a force driving the uplift of waters from the oceans to the top of the mountains. He again referred to the similarity between the earth and the human body, saying that *"if the body of the earth were not like that of a man, it would be impossible that the waters of the sea, being so much lower than the mountains, could by their nature rise up to the summits of these mountains. Hence it is to be believed that the same cause which keeps the blood at the top of the head in man keeps the water at the summits of the mountains"* (Codex Paris, A).

Then, Leonardo refers for the first time to the idea of a heat source, causing the rise of the blood inside the human body, as well as the waters inside the earth: *"I say that just as the natural heat of the blood in the veins keeps it in the head of man – for when the man is dead the cold blood sinks to the lower parts – and when the sun is hot on the head of a man the blood increases and rises so much, with other humours, that by pressure in the veins pains in the head are often caused; in the same way veins ramify through the body of the earth, and by the natural heat which is distributed throughout the containing body, the water is raised through the veins to the tops of mountains"* (Codex Paris, A). For Leonardo, it was a combination of the sun's heat and that of what he called the element of fire, which causes the rise of water inside the earth: *"this water, which passes through a closed conduit inside the body of the mountain like a dead thing, cannot come forth from its low place unless it is warmed by the vital heat of the spring time. Again, the heat of the element of fire and, by day, the heat of the sun, have power to draw forth the moisture of the low parts of the mountains and to draw them up, in the same way as it draws the clouds and collects their moisture from the bed of the sea"* (Codex Paris, A).

Here, Leonardo identifies the sun as a major source of energy, and although he recognises the crucial role of solar radiation in atmospheric moisture circulation *"drawing the clouds and collecting the moisture from the sea"* (Codex Paris, A), he fails to see that this is the key process driving the hydrological cycle. In fact, he wrote that *"in the whole universe there is nowhere to be seen a body of greater magnitude and power than the sun"*. He provided a very detailed argumentation in order to *"prove that the sun is hot by nature and not by virtue"*, writing *"this is abundantly proved by the radiance of the solar body on which the human eye cannot dwell and besides this no less manifestly by the rays reflected from a concave mirror, which – when they strike the eye with such splendour that the eye cannot bear them – have a brilliancy equal to the sun in its own place. And that this is true I prove by the fact that if the mirror has its concavity formed exactly as is requisite for the collecting and reflecting of these rays, no created being could endure the heat that strikes from the reflected rays of such a mirror*" (Codex Paris, G).

In a sense, Leonardo over-estimated the sun's importance, saying that "*its light gives light to all the celestial bodies which are distributed throughout the universe ... and there is no other centre of heat and light in the universe*". This last statement is of course false, in the sense that the sun only lights the planets of our solar system.

At a later stage of his life, Leonardo came to the conclusion that the concept of the rising of the blood due to a heat source was erroneous. Gradually, his perception of the fluxes of the hydrological cycle became closer to today's understanding. This can be concluded from his statement that: "*in many cases one and the same thing is attracted by two strong forces, namely Necessity and Potency. Water falls in the form of rain; the earth absorbs it from the necessity for moisture; and the sun evaporates it, not from necessity, but by its power*". His perception of the never-ending water cycle was further underlined when he wrote "*The water you touch in a river is the last of that which has passed, and the first of that which is coming. Thus it is with time present*". This last observation is a beautiful and poetic way of describing the irreversible time-dimension in hydrology.

Leonardo's understanding of the formation of clouds and rainfall evolved considerably over time. At an early stage, he thought clouds were generated by mist created by the waves of the sea crushing against the shores. This observation is not so strange if we realise that the land–sea interface often coincides with a zone of cloud formation due to local lifting.

According to Leonardo, the varying intensity of the rain falling from the clouds during a storm was caused by the varying strength of the wind which, with increasing speed, tends to concentrate the drops: "*But the rain which falls through the atmosphere being driven and tossed by the winds becomes rarer or denser according to the rarity or density of the winds that buffet it, and thus there is generated in the atmosphere a moisture formed of the transparent particles of the rain which is near to the eye of the spectator*" (Codex Leicester).

Leonardo eventually came up with the heat source which he supposed to be at the origin of the movement of water from the oceans towards the mountain tops: "*Where there is life there is heat, and where vital heat is, there is movement of vapour. This is proved, inasmuch as we see that the element of fire by its heat always draws to itself damp vapours and thick mists as opaque clouds, which it raises from seas as well as lakes and rivers and damp valleys; and these being drawn by degrees as far as the cold region, the first portion stops, because heat and moisture cannot exist with cold and dryness; and where the first portion stops the rest settle, and thus one portion after another being added, thick and dark clouds are formed*" (Codex Paris, A).

At this stage, Leonardo introduced the idea of the transport of energy over large distances, writing that "*they* (the clouds) *are often wafted about and borne by the winds from one region to another, where by their density they become so heavy that they fall in thick rain; and if the heat of the sun is added to the power of the element of fire, the clouds are drawn up higher still and find a greater degree of cold, in which they form ice and fall in storms of hail. Now the same heat which holds up so great a weight of water as is seen to rain from the clouds, draws them from below upwards, from the foot of the mountains, and leads and holds them within the summits of the mountains, and these, finding some fissure, issue continuously and cause rivers*" (Codex Paris, A).

Eventually, Leonardo did recognize the importance of evaporation in the hydrological cycle: "*the water finding that its element was the lordly ocean, was seized with a desire to rise above the air, and being encouraged by the element of fire and rising as a very subtle vapour, it seemed as though it were really as thin as air*".

When water vapour reaches higher altitudes, where the density of the air and temperatures are lower, condensation takes place, as Leonardo correctly wrote: "*having risen very high, it* (the water) *reached the air that was still more rare and cold, where the fire forsook it, and the minute particles, being brought together, united and became heavy; whence its height deserting it, it betook itself to flight and it fell from the sky, and was drunk up by the dry earth, where, being imprisoned for a long time, it did penance for its sin*".

Once he began to realize the importance of evaporation of water as being the engine for the uplift of water towards the tops of the mountains, Leonardo gave numerous illustrations of his conceptualisation of the hydrological cycle:

- "*All the elements will be seen mixed together in a great whirling mass, now borne towards the centre of the world, now towards the sky; and now furiously rushing from the South towards the frozen North, and sometimes from the East towards the West, and then again from this hemisphere to the other*". In these statements, Leonardo emphasised the fact that both water and soil can be transported over very large distances through the carrying winds.
- "*A great part of the sea will fly towards heaven and for a long time will not return*". Here he insists on the enormous quantities of water that are carried as water vapour during the atmospheric part of the hydrological cycle.
- "*The waters of the sea will rise above the high peaks of the mountains towards heaven and fall again on to the dwellings of men*". While still referring to the atmospheric part of the hydrological cycle, here Leonardo gives a short but accurate description of the water cycle. Through evaporation, water is transported over large distances towards the mountainous areas, where through the condensation process it will ultimately fall towards the earth and feed the rivers and man-made canals.
- "*The greatest mountains, even those which are remote from the sea shore, will drive the sea from its place*". He gave much attention to the erosion of soils and the sedimentation in estuaries. He supposed sedimentation was capable of forcing seas to retire and migrate to other locations: "*this is by rivers which carry the earth they wash away from the mountains and bear it to the shore; and where the earth comes the sea must retire*".
- According to Leonardo, the rain that fell to the earth would remain stored for a certain amount of time in ponds, lakes and oceans: "*The water dropped from the clouds still in motion on the flanks of mountains will lie still for a long period of time without any motion whatever; and this will happen in many and diverse lands*". In the same sense, he explained the formation of lakes and seas: "*all the lakes and all the gulfs of the sea and all inland seas are due to rivers which distribute their waters into them, and from impediments in their downfall into the Mediterranean Sea, which divides Africa from Europe and Europe from Asia by means of the Nile and the Don which pour their waters into it*".

At this stage of his conceptualisation of the water cycle, Leonardo had identified certain key processes, their sequence and interaction. His difficulties in finding the correct interrelations between processes are illustrated by his writings on the impact of evaporation on the salinity of water.

5.3.4 The water cycle: salinity of the sea water

When trying to explain the salinity of sea water Leonardo referred to the writings of Pliny, saying that: "*the water of the sea is salt because the heat of the sun dries up the moisture and drinks it up; and this gives to the wide stretching sea the savour of salt*" (Codex Paris, G). But, Leonardo did not accept this theory since "*if the salinity of the sea were caused by the heat of the sun, there could be no doubt that lakes, pools and marshes would be so much the more salty, as their waters have less motion and are of less depth; but experience shows us, on the contrary, that these lakes have their waters quite free from salt*" (Codex Paris, G).

Leonardo again referred to Pliny who thought that "*this salinity might originate, because all the sweet and subtle portions which the heat attracts easily being taken away, the more bitter and coarser part will remain, and thus the water on the surface is fresher than at the bottom*". Herein, Leonardo finds a contradiction, saying that if Pliny was right "*the same thing would happen in marshes and other waters, which are dried up by the heat*".

Several reasons are given for negating the statement that "*the salinity of the sea is the sweat of the earth*". He stated that if this theory were true, "*all the springs of water which penetrate through the earth, would then be saline*".

Instead, Leonardo suggested that "*the salinity of the sea must proceed from the many springs of water which, as they penetrate into the earth, find mines of salt and these they dissolve in part, and carry with them to the ocean and the other seas, whence the clouds, the begetters of rivers, never carry it up*". Based on this theory, he correctly concluded that: "*the sea would be more saline in our times than ever it was at any time*". He also believed that "*the waters of the saline sea are fresh at the greatest depths*".

Interestingly, Leonardo used his concept of tectonics in order to explain that his theory of sea salinity could never lead to "*that in infinite time the sea would dry up or congeal into salt*". He concluded that "*this salt is restored to the earth by the setting free of that part of the earth which rises out of the sea with the salt it has acquired, and the rivers return it to the earth*" (Codex Paris, G).

5.3.5 Leonardo: the first experimental hydrologist?

As we have seen in the previous chapters, Leonardo's highly developed observational skills allowed him to understand some of the most important hydrological and geomorphological processes. This knowledge allowed him to develop his concepts on the water and geomorphological cycles.

Although many of his conceptualisations were based on the insights that he had gained through simple observation, Leonardo knew about the potential of well-designed instruments to improve the understanding of physical processes. We have already discussed, in Section 3.3, the meteorological instruments that Leonardo developed. In his notes there are also designs of what contemporary biographers of Leonardo have interpreted as the fundamentals of an optical telescope.

To better understand hydrological processes, Leonardo attempted to find new ways for observing and understanding the pathways followed by the rain once it reaches the ground. In the end, his quest comes close to that of all modern field hydrologists, who want to understand where the water goes when it rains, which pathways it follows and how long it resides in a catchment. In this context, it is amazing to see how Leonardo proposed an artificial tracer to study these pathways. He suggested throwing sawdust into a river, so "*to observe where the water is driven back after having struck the river banks ... and where another water joins or separates*" (Codex Paris, K). There is little difference here to the use of artificial tracers used in modern hydrological process studies.

While relying on his observational skills, Leonardo presented several ideas that are close to those employed in modern experimental hydrology. For example, he came close to a concept still used today for the identification of the origin of water in river basins: the identification of source areas that feed a river. Leonardo described water as "*looking as if it was using as many natures as the number of regions it is travelling through*" (Codex Atlanticus), meaning essentially that by dissolving elements that are characteristic of a given region, water takes the physical and chemical signature of this same region. Since, as an artist and a writer, he liked to give figurative examples, Leonardo described water as behaving "*like a mirror that changes its colour according to that of the object that is put in front of it, modifying itself according to the place where it passes by: clean, harmful, laxative, astringent, sulfurous, salty, incarnadine* (bloodlike)*, disastrous, raging, fiery, red, yellow, green, black, blue, clayey, subtle*" (Codex Arundel). He also noted that "*in a never ending change, sometimes of the site, sometimes of the colour, it* (the water) *soaks, smells and flavours, and sometimes it retains essences or new properties, sometimes fatal, sometimes life-spending, it dissipates in the air, and sometimes it suffers from the heat that attracts it towards the colder regions, where this same heat guided it and is then contained*" (Codex Arundel).

In a way, these explanations come close to one of the fundamental aspects of contemporary experimental hydrology, relying on the decisive influence of the physiographical characteristics of basins on runoff generating processes (Buttle, 1998; Beven, 2001). It is these characteristics that cause the spatial organisation of a catchment, which Leonardo indirectly but in a beautiful and poetic way described as being at the origin of the different colours, smells and tastes of flowing waters. According to the spatial distribution of flow paths, the patterns of water storage, as well as the residence time of the water in a river basin, the water takes on the physical and chemical characteristics belonging to these factors.

This aspect is of such importance to modern hydrology that it deserves some more attention, if only to better understand the relevance of Leonardo's description of water as "*modifying itself according to the place where it passed by*" (Codex Atlanticus).

The great spatial variability in runoff producing processes is the root cause of a serious stagnation of progress in hillslope hydrology over the past 30 years (Weiler & McDonnell, 2004). The complexity of the runoff producing processes lead to a steady increase of studies focusing on the physical processes behind runoff production in many experimental micro-scale research catchments. This is in strong contrast with water management concerns that call for more understanding of hydrological processes at larger scales (Soulsby *et al.*, 2006). Today's needs for the predictions of hydrological and ecological impacts of land use and climate change must account for the spatial variability that characterises nature (Grayson & Blöschl, 2000). These research priorities may seem very different to those that prevailed during the early 16th century. However, Leonardo was also driven by his motivation to better understand hydrological and hydraulic processes so as to optimise the efficiency of the measures he designed for flood protection, as well as for agricultural use of water. In fact, the motivation for water studies has remained the same over the centuries, namely to enhance our control of water resources.

Continuous observation of water levels and discharge are fundamental to modern hydrology. From Leonardo's writings, it is apparent that he invented technical solutions for the evaluation of river discharge through measurements of riverbed dimensions and flow velocity.

Leonardo wrote about "*the measurement of water and its variations*". He noted that "*the amount* (flow) *of water that passes through a given opening varies in different ways*" depending on whether (Codex Paris, F):

1. *the surface of the water is located more or less above the opening through which it passes,*
2. *the water passes with more or less speed over the dike in which the above opening has been built,*
3. *the slope of the surface over which the water passes through the opening is more or less steep,*
4. *the inclination of the sides of this opening is more or less important,*
5. *the size of the bottom of the opening* (is small or large),
6. *the shape of the opening is a circle, a square, a rectangle or elongated,*
7. *the opening is placed in a more or less oblique way with respect to the length of the dike,*
8. *the opening is placed in a more or less oblique way with respect to the height of the dike,*
9. *the opening is located in the concave or convex part of the dike,*
10. *the canal is more or less broad,*
11. *at the highest part of the canal the speed of water is greatest above the opening,*
12. *the riverbed is irregular* (higher or lower) *next to this opening,*
13. *the water passing through the opening is exposed to the wind,*
14. *the water that falls through the air after having passed through the opening is circumscribed* (in a confined jet) *in all directions, except to the front,*
15. *the water falls into a pool of greater or less depth,*
16. *the water falls over a distance of greater less length,*
17. *the roughness of the sides of the pipe through which the water passes, or if it is straight or curved.*

Although no. 15 is wrong, most of the 17 points are still valid today, when it comes to designing a discharge measuring device (a weir) in a stream. Such a structure can be used for both discharge measurement and water level control. Common devices in hydrology are thin-plate weirs, thin-plate notch weirs (rectangular, triangular, trapezoidal or circular notch cut in a thin plate), broad-crested weirs, triangular-profile weirs, as well as flat-V weirs.

In his notes, Leonardo gave a detailed description of an experimental pipe, which would have "*one side made of glass and the remaining parts made of wood*" (Codex Paris, I). Papyrus fragments added to the water circulating in the pipe were supposed to help to identify the eddies of the water inside the pipe. This type of device is similar to those used in modern hydraulic laboratories.

Leonardo also tried to think of ways and methods to estimate "*how much water travels in one hour*", and described a method to measure flow velocity in a canal. He knew about the importance of precise measurements of time for obtaining reliable velocity measurements. They allow for the calculation of the time it takes a floating object to travel a given distance on the surface of flowing water.

Knowing that Leonardo made flow velocity measurements in canals, it seems obvious that he would also use them to estimate discharge. However, this cannot be clearly inferred from his writings. He stated that he considered his flow velocity measurements as not suited for rivers, because their flow conditions were too heterogeneous between the riverbed and the surface of the flowing water. In his notes, Leonardo refers to "*the different velocities of flowing waters, from the bottom to the surface*", "*the different currents at the surface of flowing waters*", "*the various water depths in rivers*", "*where the flow velocity is high at the surface and low close to the bottom*", or "*where the flow velocity is low at the surface and high close to the bottom*" (Codex Paris, F). These statements are yet another example of Leonardo's outstanding observational skills. Perhaps it is because of these observations that he was very careful in applying his flow velocity measurement method to rivers. The differences in flow conditions between open canals and natural riverbeds were simply too large. Realising the high variability of flow velocity from the riverbed to the surface in rivers, he presumably concluded that the velocity measurements made with a floating device were just not reliable enough to make accurate discharge estimations in natural rivers.

Leonardo also designed a specific device that was to be used for the measurement of travelling speeds of both wind and water. This device consists of a vertical plate with two holes (the lower one being five times bigger than the upper one), above which two perforated cones are fixed. In front of the two cones is installed a propeller that has its horizontal shaft linked to a weight through a rope (Fig. 5.3).

According to the travelling speed of the wind or water through the cones, the weight will be lifted more or less high. Leonardo included two cones, so as to incorporate the possibility to evaluate the accuracy of the measurement. Indeed, the operator was supposed to make two individual speed measurements: one with the upper cone, and one with the lower one. Since there is a ratio of 1 to 5 between the basal areas of the two cones, the same ratio also has to apply to the two observed lifting heights of the weight attached to the rope. This illustrates that Leonardo had clearly identified the importance of considering uncertainties when performing measurements. The identification and quantitification of uncertainties still remains today as one of the top priorities in hydrological sciences.

In modern hydrology, discharge measurements are generally still considered inaccurate. In small watercourses, discharge estimates are regarded as being reasonably accurate when calculated via the individual level–discharge relations that are specific

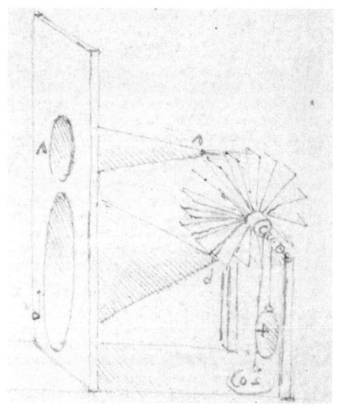

Fig. 5.3 Device for the measurement of the travelling speed of wind or water (Codex Arundel). Copyright © The British Library Board. All Rights Reserved, Arundel 263, f.241.

to the individual types of weirs. On large rivers, however, both water level measurements, as well as discharge measurements, can contain considerable errors, depending on factors such as flow variability, riverbed instability, etc. The difficulty Leonardo encountered in estimating average flow velocity is still a challenge today. Some techniques are based on individual point velocity measurements at various depths and locations between two opposite riverbanks. The individual flow velocity measurements are then integrated to obtain the overall flow velocity at a given moment on a given river section. The first current meters, which estimate flow velocity with propellers, were designed and built in the second half of the 19th century; e.g. that by Fteley & Stearns in 1873 (Herschy & Fairbridge, 1998). For decades flow velocity measurements were based on these mechanical devices, but new instruments based on electromagnetic principles and ultrasonics are now being used increasingly.

Leonardo's approach to flow measurement in open water comes close to that of modern hydrology. It is mainly the technological advances made in recent times that have allowed significant progress in experimental hydrology. Given that Leonardo asked himself many of the fundamental questions still valid today regarding where the water comes from and where it goes, he can be truly seen as an experimental hydrologist. His views, concepts, designs and solutions were well ahead of his time.

5.3.6 On the origin of tides and the link with Leonardo's theory of the hydrological cycle

In his writings on the hydrological cycle, Leonardo discussed the origin of the ebb and flow caused by the tide, questionning: *"whether flood and ebb are caused by the moon or the sun, or by the breathing of this terrestrial machine"* (Codex Leicester).

Erroneously, Leonardo considered the tides to be synchronous, so that the rise and fall would occur simultaneously anywhere on earth. His eventual explanation for this was that they only *"seem to vary in time because the days do not begin at the same time throughout the universe; in such a way that when it is midday in our hemisphere, it is midnight in the opposite hemisphere"* (Codex Leicester). This simultaneous *"swelling and diminution in the height of the seas"* was supposed to be due to the withdrawal of the waters from the bottom of the oceans *"into the fissures which start from the depths of the sea and which branch inside the body of the earth"*. Here, Leonardo's concept of a subterranean network of conduits comes in again as it is ultimately *"corresponding to the sources of rivers, which are constantly taking from the bottom of the sea the water which has flowed into it"*. As a consequence, the sources were supposed to be fed through *"a sea of water which is incessantly being drawn off from the surface of the sea"*.

Leonardo provided several arguments to prove that the moon could not be at the origin of the ebb and flow. Referring to the Mediterranean Sea, he basically considered that *"if such a superabundance of water had to pass through the Straits of Gibraltar in running behind the moon, the rush of the water through that strait would be so great, and would rise to such a height, that beyond the straits it would for many miles rush so violently into the ocean as to cause floods and tremendous seething, so that it would be impossible to pass through"*.

Leonardo was well aware of the fact that the magnitude of the tide varied considerably from one geographical region to another, writing *"that flood and ebb are not general; for on the shore at Genoa there is none, at Venice 2 braccia, between England and Flanders 18 braccia"* (Codex Leicester). And, *"in the West, near to Flanders, the sea rises and decreases every six hours about 20 braccia, and 22 when the moon is in its favour; but 20 braccia is the general rule, and this rule, as it is evident, cannot have the moon for its cause"*. Leonardo suggested that *"this variation in the increase and decrease of the sea every six hours may arise from the damming up of the waters, which are poured into the Mediterranean by the quantity of rivers from Africa, Asia and Europe, which flow into that sea, and the waters which are given to it by those rivers; it pours them to the ocean through the Straits of Gibraltar. That ocean extends to the island of England and others farther North, and it becomes dammed up and kept high in various gulfs. These, being seas of which the surface is remote from the centre of the earth, have acquired a weight, which as it is greater than the force of the incoming waters which cause it, gives this water an impetus in the contrary direction to that in which it came and it is borne back to meet the waters coming out of the straits; and this it does most against the straits of Gibraltar; these, so long as this goes on, remain dammed up and all the water which is poured out meanwhile by the aforementioned rivers, is pent up* (in the Mediterranean Sea); *and this might be assigned as the cause of its flow and ebb"* (Codex Leicester). Here it is noteworthy that Leonardo came close to understanding the standing wave character of the tide. The fact that the tidal range varies so strongly over the North Sea (and also over the Mediterranean, although less strongly) is indeed caused by interference of the shorelines with the tidal wave, which causes strong amplification in some areas and almost no tide in others. His idea of water backing up and receiving an "***impetus in the contrary direction***" comes close to the refracted wave that, in combination with the incoming wave, causes a standing wave. Without the help of wave equations this phenomenon is difficult to explain or to understand from simple observations.

Leonardo also described the propagation of the tide in estuaries and water courses subject to the tide. Although in the following text he refers to a mountain stream, which of course is not subject to ocean tides: "***In many places there are streams of water which swell for six hours and ebb for six hours; and I, for my part, have seen one above the lake of Como called Fonte Pliniana, which increases and ebbs, as I have said, in such a way as to turn the stones of two mills; and when it fails it falls so low that it is like looking at water in a deep pit***" (Codex Leicester). He furthermore wrote that "***About eight miles above Como is the Pliniana, which increases and ebbs every six hours, and its swell supplies water for two mills; and its ebbing makes the spring dry up; two miles higher up there is Nesso, a place where a river falls with great violence into a vast rift in the mountain***" (Codex Atlanticus).

Leonardo here refers to a periodic spring, previously described by Pliny the Elder, which indeed ebbs and flows several times a day. Contrary to the statements of Pliny the Elder and Leonardo, the ebb and flow actually alternate at irregular intervals. The spring can still be visited today near the Villa Pliniana, built in 1577 near the village of Torno. Today, geologists assume that the intermittent flow is due to the hydraulic principle of a siphon.

6 THE STUDY OF WATER IN MOTION

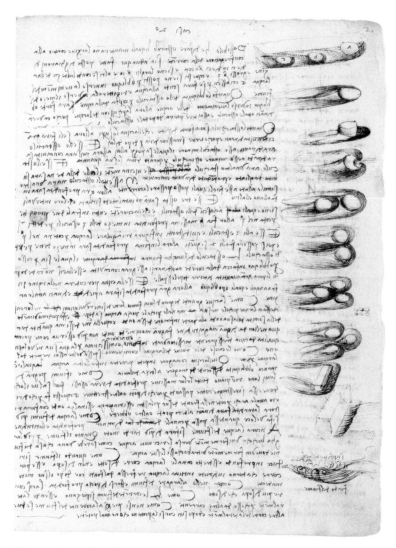

Discussion of water currents.
From Sheet Discussing Water Currents, Codex Leicester. © Seth Joel/CORBIS.

6.1   LEONARDO'S MOTIVATION FOR STUDYING MOVING WATERS

Leonardo spent a considerable effort on the study of moving waters, largely because he wanted to find new and efficient ways to protect populations from devastating floods.

This is perfectly illustrated by his numerous detailed descriptions and drawings of disasters related to devastating floods (Fig. 6.1): "***Then the ruins of the high buildings in these cities will throw up a great dust, rising up in shape like smoke or wreathed clouds against the falling rain; but the swollen waters will sweep round the pool which contains them striking in eddying whirlpools against the different obstacles, and leaping into the air in muddy foam; then, falling back, the beaten water will***

Fig. 6.1 Drawing showing a severe storm over a city in a mountainous area. The Royal Collection © 2008 Her Majesty Queen Elizabeth II.

again be dashed into the air. And the whirling waves which fly from the place of concussion, and whose impetus moves them across other eddies going in a contrary direction, after their recoil will be tossed up into the air but without dashing off from the surface. Where the water discharges from the pool the spent waves will be seen spreading out towards the outlet; and there falling or pouring through the air and gaining weight and impetus they will strike on the water below piercing it and rushing furiously to reach its depth; from which being thrown back it returns to the surface of the lake, carrying up the air that was submerged with it; and this remains at the outlet in foam mingled with logs of wood and other matters lighter than water. Round these again are formed the beginnings of waves which increase the more in circumference as they acquire more movement; and this movement rises less high in proportion as they acquire a broader base and thus they are less conspicuous as they die away. But if these waves rebound from various objects they then return in direct opposition to the others following them, observing the same law of increase in their curve as they have already acquired in the movement they started with".

As an artist, Leonardo was literally fascinated by the movements of water. The above description of a deluge provides the reader with so many details that it comes close to looking at one of Leonardo's paintings. He was probably one of the first artists to devote a lot of effort to the correct representation of motion, whether of living or inanimate bodies.

In a letter to one of his friends, Leonardo gave another very detailed description of a natural disaster (Codex Atlanticus), of which the location is not clearly known (probably it occurred in the Alps): *"Having many times rejoiced with you by letters over your prosperous fortunes, I know now that, as a friend you will be sad with me over the miserable state in which I find myself; and this is, that during the last few days I have been in so much trouble, fear, peril and loss, besides the miseries of the people here, that we have been envious of the dead; and certainly I do not believe that since the elements by their separation reduced the vast chaos to order, they have ever combined their force and fury to do so much mischief to man. As far as regards us here, what we have seen and gone through is such that I could not imagine that things could ever rise to such an amount of mischief, as we experienced in the space of ten hours. In the first place we were assailed and attacked by the violence and fury of the winds; to this was added the falling of great mountains of snow which filled up all this valley, thus destroying a great part of our city.*

And not content with this, the tempest sent a sudden flood of water to submerge all the low part of this city; added to which there came a sudden rain, or rather a ruinous torrent and flood of water, sand, mud, and stones, entangled with roots, and stems and fragments of various trees; and every kind of thing flying through the air fell upon us; finally a great fire broke out, not brought by the wind, but carried as it would seem, by ten thousand devils, which completely burnt up all this neighbourhood and it has not yet ceased. And those few who remain unhurt are in such dejection and such terror that they hardly have courage to speak to each other, as if they were stunned. Having abandoned all our business, we stay here together in the ruins of some churches, men and women mingled together, small and great, just like herds of goats. The neighbours out of pity succoured us with victuals, and they had previously been our enemies. And if it had not been for certain people who succoured us with victuals, all would have died of hunger. Now you see the state we are in. And all these evils are as nothing compared with those which are promised to us shortly.

I know that as a friend you will grieve for my misfortunes, as I, in former letters have shown my joy at your prosperity ...".

Interestingly, Leonardo identified three different causes acting simultaneously during major floods in the river systems connected to the Mediterranean Sea: *"On the shores of the Mediterranean 300 rivers flow ... And this sea is 3000 miles long. Many times has the increase of its waters, heaped up by their backward flow and the blowing of the West winds, caused the overflow of the Nile and of the rivers which flow out through the Black Sea, and have so much raised the seas that they have spread with vast floods over many countries. And these floods take place at the time when the sun melts the snows on the high mountains of Ethiopia that rise up into the cold regions of the air; and in the same way the approach of the sun acts on the mountains of Sarmatia in Asia and on those in Europe; so that the gathering together of these three things are, and always have been, the cause of tremendous floods: that is, the return flow of the sea with the West wind and the melting of the snows. So every river will overflow in Syria, in Samaria, in Judea between Sinai and the Lebanon, and in the rest of Syria between the Lebanon and the Taurus mountains, and in Cilicia, in the Armenian mountains, and in Pamphilia and in Lycia within the hills, and in Egypt as far as the Atlas mountains"* (Codex Leicester).

Leonardo erroneously thought that the overtopping of rivers such as the Nile was the result of a combination of sea level rise due to westerly winds, a return flow from the Mediterranean Sea and the rising discharge due to snow cover melting in the mountainous headwaters. In fact, the hydrology of the Nile is extremely complex. While the White Nile has a relatively stable discharge throughout the year, the Blue Nile has a high seasonal variability, induced by the successive wet and dry seasons. With more than two-thirds of the total annual flow of the Nile stemming from the Blue Nile, the annual discharge of the total Nile varied by a factor of 15 over the year (maximum discharge in August/September and minimum discharge in April/May), before the construction of dams during the 20th century.

6.2 TRAINING RIVERS, OR HOW TO RAISE THE ECONOMIC VALUE OF RIVERS

Driven by his willingness to protect populations from the devastating force of floods, Leonardo knew that he first had to understand the laws that determine the flow of waters in riverbeds and flood plains.

Leonardo gave much attention to the River Arno. Most probably his interest in the study of water was closely linked to the devastating floods that the Arno caused in 1456 and 1466. One of his ideas to fight the devastating force of water was to transform the River Arno into a navigable canal from Pisa to Florence. He also referred to the economic benefits of such a plan, noting that *"by guiding the Arno above and below a treasure will be found in each acre of ground by whomsoever will"* (Codex Atlanticus).

He obviously spent a lot of time observing the River Arno, trying to understand its extreme hydrological and hydraulic behaviour. These observations gave him a clear and correct view of some of the basic principles of river morphology. In one of his notes, Leonardo stated that "*they do not know why the Arno will never remain in a channel. It is because the rivers which flow into it deposit earth where they enter, and wear it away on the opposite side, bending the river in that direction*" (Windsor Drawings). Here he refers to the meanders of the river characterised by a cliff on the outside (where flow speed is highest) and a gentle slope on the inner side of the bend (where flow speed is lowest).

Leonardo's idea was to create a canal that would on the one hand stabilise the course of the river, and on the other hand allow the regulation of the discharge of the Arno. The regulation of the discharge in the canal was to be handled via a complex system of sluices. He predicted additional economic benefits for "***Prato, Pistoia and Pisa, as well as Florence, that will gain two hundred thousand ducats a year, and will lend a hand and money to this useful work***" (Codex Atlanticus).

In the Mediterranean climate, the control of water for irrigation had been a focal point of interest for thousands of years. Leonardo was well aware of this and tried to figure out strategies to limit water losses to a minimum in the canal systems. The construction of canal systems had already been initiated as early as the 12th century to link the city of Milan to the rivers Ticino and Adda (two left-bank tributaries of the Po River, thus ultimately also giving the city of Milan access to the Mediterranean Sea), both for irrigation purposes and the transport of economic goods. The construction of the so-called *Naviglio Martesana* started in 1457 and created a link between the Adda River and the city of Milan. Its construction was of greatest strategic importance for the city of Milan in its long military and economic competition with the city of Venice. In only 8 years, the construction of the 38 km of the canal of Martesana was completed. Being involved at a later stage in studies that were supposed to diminish some of the negative side-effects of its construction, he wrote that "***by making the canal of Martesana, the water of the Adda is greatly diminished by its distribution over many districts for the irrigation of the fields***" (Codex Paris, F). As a remedy, Leonardo suggested "***to make several little channels, since the water drunk up by the earth is of no more use to any one, nor mischief neither, because it is taken from no one; and by making these channels, the water which before was lost returns again and is once more serviceable and useful to men***" (Codex Paris, F).

Another example of his observations concerns the confluence between the Arno and the Mensola in a very particular situation, i.e. where the discharge of the Arno is much less than that of the Mensola. Leonardo drew a detailed sketch of this configuration, showing the eddy made at the confluence (Fig. 6.2).

More generally, Leonardo deduced from his observations of confluences that "***When a smaller river pours its waters into a larger one, and that larger one flows from the opposite direction, the course of the smaller river will bend up against the approach of the larger river; and this happens because, when the larger river fills up all its bed with water, it makes an eddy in front of the mouth of the other river, and so carries the water poured in by the smaller river with its own. When the smaller river pours its waters into the larger one, which runs across the current at the mouth of the smaller river, its waters will bend with the downward movement of the larger river***" (Codex Leicester).

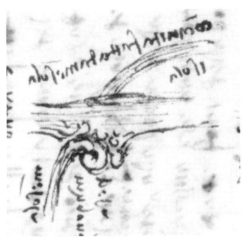

Fig. 6.2 The eddy made by the Mensola, when the Arno is low and the Mensola full. Detail from Sheet Discussing the Confluence of Rivers, Codex Leicester. © Seth Joel/CORBIS.

For the implementation of his plans to canalise the Arno, Leonardo gave clear instructions: "*the river which is to be turned from one place to another must be coaxed and not treated roughly or with violence; and to do this a sort of floodgate should be made in the river, and then lower down one in front of it and in like manner a third, fourth and fifth, so that the river may discharge itself into the channel given to it, or that by this means it may be diverted from the place it has damaged, as was done in Flanders, as I was told by Niccolo di Forsore*" (Codex Leicester). Here he referred to similar constructions in Flanders, known for the numerous hydraulic devices built for drainage and flood protection over several centuries.

Regarding the canal at Florence, Leonardo wrote: "*Sluices should be made in the valley of La Chiana at Arezzo, so that when, in the summer, the Arno lacks water, the canal may not remain dry: and let this canal be 20 braccia wide at the bottom, and at the top 30, and 2 braccia deep, or 4, so that two of these braccia may flow to the mills and the meadows, which will benefit the country*" (Codex Atlanticus). What Leonardo proposes is to build according to natural laws, or as we say nowadays: to build with nature (and not against it). If canals are designed as close as possible to the natural shape and slope of rivers, the maintenance required is expected to be minimal, and the chance that the construction will last for a longer period of time is greatest. Any hard engineering intervention, such as locks or bank protection works requires constant maintenance and high operational costs.

Furthermore, he provided detailed information on the costs of the works to be executed: "*And know that in digging this canal where it is 4 braccia deep, it will cost 4 dinari per square braccio; for twice the depth 6 dinari, if you are making 4 braccia and there are but 2 banks*". He even proposed different options, rated according to their costs as well as their benefits.

Leonardo designed river training works and dimensioned canals on the basis of his empirical findings, primarily through observations. As we have previously seen in Section 5.3.5, Leonardo presumably knew how to estimate a discharge, i.e. by multiplying velocity with the cross-sectional area. On the relationship between riverbed dimensions and discharge, he wrote "*A river of equal depth runs with greater speed in a narrow space than in a wide one, in proportion to the difference between the wider and the narrower one. This proposition is clearly proved by reason confirmed by experiment. Supposing that through a channel one mile wide there flows one mile in length of water; where the river is five miles wide each of the 5 square miles will require 1/5 of itself to be equal to the square mile of water required in the body of water, and where the river is 3 miles wide each of these square miles will require the third of its volume to make up the amount of the square mile of the narrow part*".

Leonardo's tremendous observational skills allowed him to often draw correct conclusions related to river hydraulics. This is illustrated by the technical solution he suggests to counteract riverbed erosion caused by collapsing cliffs: "*Again if the lowest part of the bank which lies across the current of the waters is made in deep and wide steps, after the manner of stairs, the waters which, in their course usually fall perpendicularly from the top of such a place to the bottom, and wear away the foundations of this bank can no longer descend with a blow of too great a force; and I find the example of this in the stairs down which the water falls in the fields at Sforzesca at Vigevano over which the running water falls over a height of 50 braccia*" (Codex Leicester).

At a later stage of his life, when staying in Amboise, France, Leonardo studied the River Loire. His notes show many details of the aspects he paid attention to while observing the flow of the river (Fig. 6.3): "*This is the river that passes through Amboise; it passes at a–b/c–d, and when it has passed the bridge it turns back, against the original current, by the channel d–e/b–f in contact with the bank which lies between the two contrary currents of the said river, a–b/c–d, and d–e/b–f. It then turns down again by the channel f–l/g–h/n–m, and reunites with the river from which it was at first separated, which passes by k–n, which makes k–m/r–t. But when the river is very full it flows all in one channel passing over the bank b–d* (Codex Arundel)".

During these observations, Leonardo evidently noticed a splitting of the flow of the River Loire due to two islands located in the middle of the riverbed. Apparently, he observed a return flow between these two islands, in the opposite direction to the main downstream flow of the Loire. From his notes it is difficult to decipher whether the observed flow patterns were only restricted to the surface, or if they apply to the entire depth of the river. He made a clear distinction between the flow patterns during low and high flow conditions, the latter causing a more homogeneous downward flow over one of the islands. The description of the flow patterns of the Loire near Amboise is a beautiful illustration of Leonardo's attempts to understand fluid dynamics through observation.

While there is no doubt that Leonardo's motivation for the design and implementation of new riverbeds and canal networks was largely driven by his wish to fight the devastating forces of floods and the redistribution of water for agricultural purposes, it appears that, at least to some extent, his work was also relevant for military purposes.

During Leonardo's time, Florence was engaged in a long and energy-consuming war against Pisa, which had declared its independence in 1494. Despite many military attempts, no significant progress was made for several years. As soon as he returned to

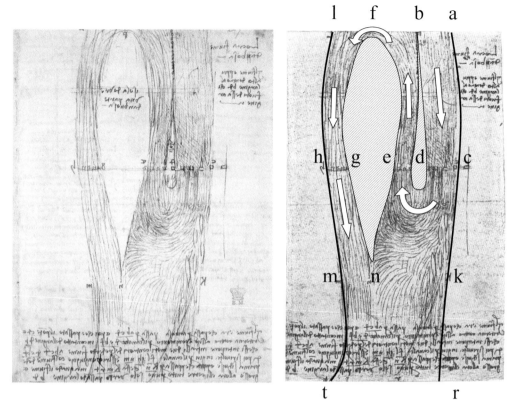

Fig. 6.3 *Hydraulics of the River Loire at Amboise. Codex Arundel. Copyright © The British Library Board. All Rights Reserved. Arundel 263, ff.264v,269.*

Florence, Leonardo worked on a proposal which consisted in diverting the course of the Arno, so as to cut off the water supply to Pisa and divert the river into the Stagno marsh. While it is not known who originally had the idea of this flow diversion, there is no doubt that Leonardo attended several planning meetings with military staff and politicians of Florence. His mission during these meetings was obviously to evaluate the feasibility of the project. Although the project appeared to be quite outlandish to many people, after more than a year of discussions it was evaluated as of great use to Florence, whether because of the diversion of the Arno to Stagno, or to present a threat to the enemy (McCurdy, 1928).

The construction of the diversion was authorised in August 1504. Thousands of ducats were to be spent on the enterprise, "*a dam was to be built across the river to stop its course, and two trenches seven braccia deep, and one twenty, the other thirty braccia wide, were to be dug to lead it to the marsh and thence to the sea*", 30 000 to 40 000 work days and 2000 men were estimated to be necessary.

However, the project completely collapsed when a flood occurred after the completion of the first trench, during which the water entered the canal and cut back to the riverbed. Since the costs of the project were literally exploding and the technical difficulties appeared much greater than anticipated, the project was stopped just two months after it had started. The abandoned trenches were subsequently filled in by the Pisans.

Although, in his notes, Leonardo defined the conditions under which a river may be diverted, there is no proof that the failure of the Arno diversion was due to his plans. There is no doubt that Leonardo spent considerable time finding the optimal solution for canal construction. He always kept in mind the technical difficulties of such

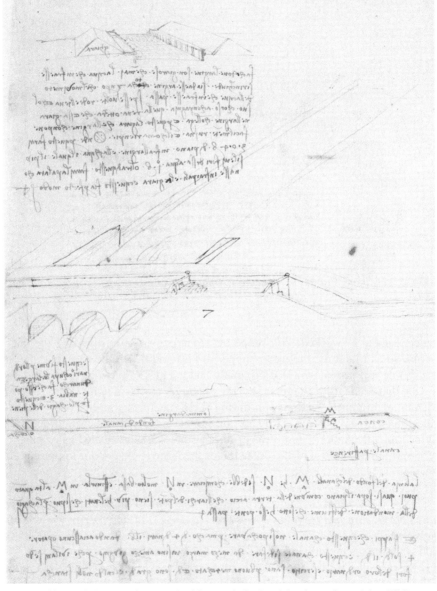

Fig. 6.4 Hatch sluice gate for a navigable canal. Codex Atlanticus. Copyrights Biblioteca Ambrosiana Auth No. Int 59/08.

projects, knowing that this would help to control the costs. For example, he aimed to use the lowest possible number of locks in a canal, arguing that their maintenance would be costly. Some of his drawings illustrate the design of a canal bridge over a river, allowing ships to pass via a system of shipping locks (Fig. 6.4). Other drawings represent the locks and their rules for raising or lowering ships during their journey through a canal (Fig. 6.5).

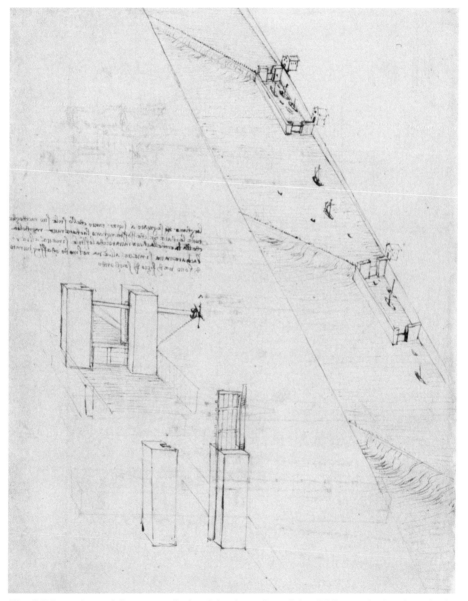

Fig. 6.5 Drop-down sluice gates. Codex Atlanticus. Copyrights Biblioteca Ambrosiana Auth No. Int 59/08.

The technique of rising and lowering ships in a river with a variable discharge had been used for more than 200 years in Lombardy, but Leonardo made substantial contributions to improving the operation of such devices.

He eventually designed a new type of gate for the shipping locks (Fig. 6.6). Operated from above at the riverbank, a hatch connected to a bolt in the lock helped to regulate the inflow and outflow of water. This in turn facilitated the regulation of the pressure on the sides of the gates and ultimately eased their opening and closing.

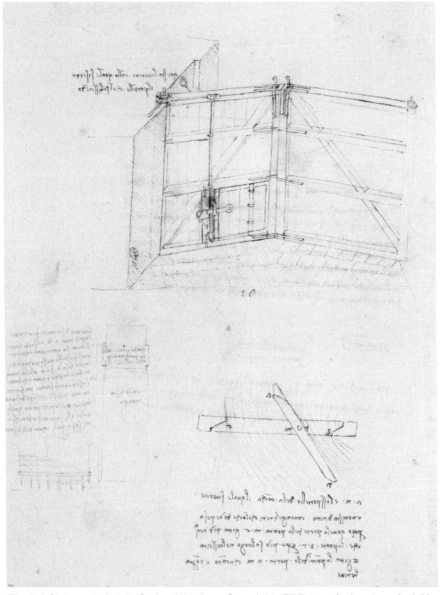

Fig. 6.6 *Sluice gate hatch. Codex Atlanticus. Copyrights Biblioteca Ambrosiana Auth No. Int 59/08.*

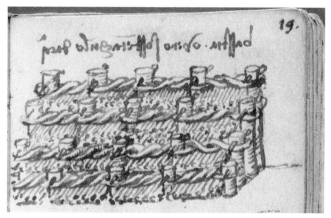

Fig. 6.7 *Canal reinforcement gabion structure. Codex Paris, Manuscript L. © RMN (Institut de France)/Gérard Blot. Paris, bibliothèque de l'Institut.*

Well aware of the pressure that falling water could exercise on the banks and the bottom of the shipping locks, Leonardo designed a special reinforcement structure for these vulnerable spots (Fig. 6.7). Where the pressure would be highest, he planned to build gabions as reinforcement. This structure consisted of rows of poles that were rammed into the banks and connected by transoms.

7 LEONARDO'S LEGACY

An important shortcoming in Leonardo's work was that he failed to make the correct assumption regarding the function of the heart in the circulation of blood. Surprisingly perhaps, for one who knew so much about the heart, including the fact that blood vessels contract and dilate, he did not see the essential link between the movement of the blood around the body and the mechanical pumping of the heart. Similarly, he did not understand the causes of the circulation of water on earth, thinking that the processes of rainfall and infiltration alone could not explain the amount of water flowing in rivers. He believed that both in the human body and in the earth "life" was sustained by heat: "*As heat brings the blood in the veins to the head of man ... for the natural heat that is in the earth, water flows in the veins on the top of the mountains*". The reason behind this is Leonardo's commitment to a grand holistic view, a rare example of his obsession becoming an impediment to his studies, eventually leading to an entirely false conclusion.

Because Leonardo was so keen on drawing comparisons between all areas of nature, he decided, as we saw in the extracts from his earlier writings, that the blood of the body was analogous to the water of the oceans, calling it "*a pool of blood, wherein the breathing lung increases and decreases*".

Only towards the end of his life did he seem to have some doubts about the analogy between the sea and blood, writing that "*the origin of the sea is opposite to the origin of blood, as the sea collects the water from all the rivers which are generated only by moisture that rises in the air; but the sea of blood is the cause of all veins*". However, the remaining fragments of his writing do not allow us to determine just how close he managed to come to the truth.

While sometimes he reached wrong conclusions, his mind was always open to new ideas, and he continuously questioned established theories, and incorporated new evidence into his interpretation of how things work, allowing him to develop theories that are close to what today we consider to be correct. His interpretation of atmospheric processes and his evolutionist theory of the movement of the earth are just a few examples of his great ideas that preceded by several centuries those of formal science today. Leonardo believed that "*the knowledge of past times and of the places on the earth is both ornament and nutriment to the human mind*". And, he was one of the first to develop the idea that geological and geographical facts could be derived from a study of the physical past; an idea we take for granted today.

However, probably the most fascinating aspect of his work, from the perspective of our time, is not what was correct and what was wrong in his vision of the world. The interesting aspect, which proves his genius, is the development of a method that was completely unique and novel in his time. This method is a collection of many elements, which he developed for the first time, and used as tools to support the creativity of his mind. The value of those elements is nowadays well established, but it is in Leonardo that for the first (and probably only) time they all co-exist together. One of those elements is the scientific approach, which would be introduced much later, during the Enlightenment. This approach allowed him to believe in what he saw, rather than in what were consolidated and affirmed theories, and to go beyond the *status quo* of his times. It also allowed him to abandon ideas that were close to his heart, such as the similarity between Man and earth, when the experimental evidence did not support them any longer. Another aspect is his real interdisciplinary view, unique in all history, which ranged from fundamental to applied science to different forms of art. This view reinforced every aspect of his ability and his knowledge. Finally, he understood the value of experiments and "data" in understanding, interpreting and predicting the behaviour of the natural system. Among other instruments, he invented a hygrometer, which could be used "*to know the qualities and thickness of the air, and when it is going to rain*".

Water, in all its aspects, attracted his attention throughout his life. When he was 14 years of age, in 1466, the River Arno burst its banks and flooded the region. This disaster, almost certainly, had an enormous impact on Leonardo's life in Anchiano, and may have played a significant role in nurturing his lifelong obsession with water.

When speaking about water he had to invent a new language, in which technical and poetic terms often co-exist together. For him, water, as part of the "*vegetative soul*" of nature, was alive and could assume different "*personalities*": "*when turbulent and destructive she goes angry, when shiny and quiet in the fresh herbs she goes joking*", and again "*when she runs and when she is quiet, when she causes life or death, when she gives and when she takes, when she feeds and when the opposite, when she is with or without salt, when with big deluges she submerges big valleys*".

In one of his notes, Leonardo described the end of the water cycle, which comes quite close to what we today think to be the final fate of our planet. Life is indeed supposed to disappear in several billion years, when the sun has come to the end of its lifetime, extinguishing all forms of life. Leonardo wrote that "*The watery element was left enclosed between the raised banks of the rivers, and the sea was seen between the uplifted earth and the surrounding air which has to envelope and enclose the complicated machine of the earth, and whose mass, standing between the water and the element of fire, remained much restricted and deprived of its indispensable moisture; the rivers will be deprived of their waters, the fruitful earth will put forth no more her light verdure; the fields will no more be decked with waving corn; all the animals, finding no fresh grass for pasture, will die and food will then be lacking to the lions and wolves and other beasts of prey, and to men who after many efforts will be compelled to abandon their life, and the human race will die out. In this way the fertile and fruitful earth will remain deserted, arid and sterile from the water being shut up in its interior, and from the activity of nature it will continue a little time to increase until the cold and subtle air being gone, it will be forced to end with the element of fire; and then its surface will be left burnt up to cinder and this will be the end of all terrestrial nature*" (Codex Arundel).

Unlike his artistic work, his scientific ideas did not have much impact in the centuries following his life. Most of his writings were lost, and others remained hidden and inaccessible in libraries and monasteries for several centuries. The discovery of Leonardo as a scientist is relatively recent. The work of recovering and decoding his writings only began at the beginning of the 19th century. His papers are difficult to read and to interpret, as they do not have the structured form of a complete work, and it is only by combining the many fragments extracted from his papers that it was possible to reconstruct his views.

The study of Leonardo's writings has revealed his tremendous observational capabilities and skills, which he was able to deploy when observing hydrological and meteorological processes. We have seen from the study of his notes that his eagerness to understand the flow paths of water was at least partially inspired by his wish to find new river management techniques, leading in the end to a higher flood protection efficiency, as well as to an improvement of water use for socio-economic purposes. In a sense, Leonardo was committed, just as modern hydrologists, to linking water science and water management (Grayson, 2005). Timeliness is always a critical aspect of research, in the sense that the right answer a year late is often useless to a water manager. A direct implication of this is that a decision on an urgent problem is always made, with or without good science! This issue is of such importance that there is rising concern about the adequacy of the education of current and future water managers, especially with respect to the rising complexity of water-related issues (Berndtsson *et al.*, 2005). Modern hydrologists have to deal with complex interrelated environmental, hydrological, ecological and socio-economic issues.

Leonardo's achievements appear even more outstanding in the light of the progress made in modern hydrology over the past decades. The development of highly sensitive sensors, connected to loggers, allows continuous monitoring of numerous hydrological and meteorological variables. Combined with hydro-chemical and isotopic investigation techniques, these devices have led to important advances in the identification of hydrological flow paths at hillslope scales. However, there is still much to be discovered concerning the incorporation of the knowledge gained on hillslopes into meso-scale basins (McDonnell, 2003), which cannot be simply summarised by assuming that headwaters are the sum of hillslopes (Uchida *et al.*, 2005).

Indeed, in 2002, the scientific community of hydrologists launched the IAHS (International Association of Hydrological Sciences) decade for Prediction in Ungauged Basins (PUB), aiming at finding new ways to bridge the gap between knowledge and concepts gained and developed in monitored basins and non-monitored basins (Clarke, 2005).

Over recent decades the development of remote sensing techniques has helped provide new potential for significant progress in hydrological sciences. However, although these techniques looked very promising at the outset of the satellite era, reality turned out to be more difficult than anticipated. Hossain & Lettenmaier (2006) recently listed a number of hydrological issues that need to be addressed before the full potential of a satellite-borne global precipitation measurement mission is likely to be exploited in the context of flood prediction. Among these issues, the assessment of surface hydrological variability in river basins, and the predictive uncertainty of hydrological models remain top priorities.

The main cause of limits to hydrological predictability lies in the nonlinearity of hydrological systems, rendering the accurate measurement of initial and boundary conditions in a hydrological system literally impossible. Blöschl & Zehe (2005) commented on this fact in an almost Leonardo-like manner: *"above all, modesty seems to be in place as to the degree to which hydrological system behaviour can be represented*".

Hydrological sciences have made tremendous progress over the past decades, certainly compared to the limited progress made between Leonardo's lifetime and the beginning of the 20th century. Nonetheless, there still remains so much to learn with respect to the question of where does the water go when it rains.

References

Ambroise, B. (1999) *La dynamique du cycle de l'eau dans un bassin-versant. Processus, facteurs, modèles*. Editions HGA, Bucarest, Romania.

Beltrando, G. & Chémery, L. (1995) *Dictionnaire du climat*. Larousse, Paris, France.

Berndtsson, R., Falkenmark, M., Lindh, G., Bahri, A. & Jinno, K. (2005) Educating the compassionate water engineer – a remedy to avoid future water management failures? *Hydrol. Sci. J.* **50**, 7–16.

Betson, R. P. (1964) What is watershed runoff? *J. Geophys. Res.* **69**, 1541–1552.

Beven, K. (2001) *Rainfall–Runoff Modelling: The Primer*. Wiley, Hoboken, New Jersey, USA.

Beven, K. J. (2006) *Streamflow Generation Processes*. IAHS Benchmark Papers in Hydrology 1. IAHS Press, Wallingford, UK.

Biswas, A. K. (1970, 1972) *History of Hydrology*. North Holland, Amsterdam, The Netherlands.

Blöschl, G. & Zehe, E. (2005) On hydrological predictability. *Hydrol. Processes* **19**, 3923–3929.

Brutsaert, W. (2008) *Hydrology – An introduction*. Cambridge University Press, Cambridge, UK.

Buttle, J. (1998) Fundamentals of small catchment hydrology. In: *Isotope Tracers in Catchment Hydrology* (ed. by C. Kendall & J. J. McDonnell), 1–43. Elsevier, Amsterdam, The Netherlands.

Clarke, R. T. (2005) The PUB decade: how should it evolve? *Hydrol. Processes* **19**, 2865–2869.

Derruau, M. (1988) *Précis de géomorphologie*. Masson, Paris, France.

Dickens, E. (2005) *The Da Vinci Notebooks by Leonardo Da Vinci*. Arcade Publishing, New York, USA.

Dooge, J. C. I. (1959) Un bilan hydrologique au 17e siècle. *La Houille Blanche* **14**(6), 799–807.

Dunne, T. & Black, R. D. (1970) Partial area contributions to storm runoff in a small New England watershed. *Water Resour. Res.* **6**, 1296–1311.

Estienne, P. & Godard, A. (1990) *Climatologie*. Colin, Paris, France.

Grayson, R. B. (2005) Delivering on the promise: linking water science and water management. *Hydrol. Processes* **19**, 2311–2313.

Grayson, R. & Blöschl, G. (2000) *Spatial Patterns in Catchment Hydrology*. Cambridge University Press, Cambridge, UK.

Herschy, R. W. & Fairbridge, R. W. (1998) *Encyclopedia of Hydrology and Water Resources*. Kluwer Academic Publishers, Dordrecht, The Netherlands.

Hewlett, J. D. & Hibbert, A. R. (1967) Factors affecting the response of small watersheds to precipitation in humid areas. In: *Proc. Int. Symp. on Forest Hydrology* (ed. by W. W. Sopper & H. W. Lull), 275–290. Pergamon, Oxford, UK.

Horton, R. E. (1933) The role of infiltration in the hydrological cycle. *Trans. Am. Geophys. Un.* **14**, 446–460.

Hossein, F. & Lettenmaier, D. P. (2006) Flood prediction in the future: recognizing hydrologic issues in anticipation of the Global Precipitation Measurement mission. *Water Resour. Res.* **42**, W11301.

Le Pichon, X. (1968) Sea floor spreading and continental drift. *J. Geophys. Res.* **73**, 3661–3697.

Lovelock, J. E. (1979) *Gaia, a New Look at Life on Earth*. Oxford University Press, Oxford, UK.

Majer, G. (2006) *Leonardo Da Vinci – Il mondo e le acque – Scritti XI*. Neri Pozza, Milano, Italy.

McCurdy, E. (1928) *The Mind of Leonardo Da Vinci*. Mineola Publishers, Dover, USA.

McCurdy, E. (1942a) *Les carnets de Léonard de Vinci – Tome 1 (*1987 Edition). Gallimard, Paris, France.

McCurdy, E. (1942b) *Les carnets de Léonard de Vinci – Tome 2* (1989 Edition). Gallimard, Paris, France.

McDonnell, J. J. (2003) Where does the water go when it rains? Moving beyond the variable source area concept of rainfall–runoff response. *Hydrol. Processes* **17**, 1869–1875.

McKenzie, D. P. & Parker, D. L. (1967) The North Pacific: an example of tectonics on a sphere. *Nature* **216**, 1276–1280.

Morgan, W. J. (1968) Rises, trenches, great faults and crustal blocks. *J. Geophys. Res.* **73**, 1959–1982.

Penman, H. L. (1963) *Vegetation and Hydrology*. Technical Communications 53. Commonwealth Bureau of Soils. Harpenden, UK.

Pfister, L. & Savenije H.H.G. (2006) Leonardo da Vinci's scriptures as benchmark papers in hydrology. Invited commentary. *Hydrol. Processes* **20**, 1653–1655.

Pinder, G. F. & Jones, J. F. (1969) Determination of the ground-water component of peak discharge from the chemistry of total runoff. *Water Resour. Res.* **5**(2), 438–445.

Richter, J.-P. (1888a) *The Notebooks of Leonardo Da Vinci* – Volume 1 (translated by R. C. Bell & E. J. Poynter). Project Gutenberg Release #4998 (January 2004).

Richter, J.-P. (1888b) *The Notebooks of Leonardo Da Vinci* – Volume 2 (translated by R. C. Bell & E. J. Poynter). Project Gutenberg Release #4999 (January 2004).

Sklash, M. G. & Farvolden, R. N. (1979) The role of groundwater in storm runoff. *J. Hydrol.* **43**, 45–65.

Soulsby, C., Tetzlaff, D., Rodgers, P., Dunn, S. & Waldron, S. (2006) Runoff processes, stream water residence times and controlling landscape characteristics in a mesoscale catchment: an initial evaluation. *J. Hydrol.* **325**, 197–221.

Strangeways, I. (2007) *Precipitation. Theory, Measurement and Distribution*. Cambridge University Press, Cambridge, UK.

Uchida, T., Asano, Y., Onda, Y. & Miyata, S. (2005) Are headwaters just the sum of hillslopes? *Hydrol. Processes* **19**, 3251–3261.

Vasari, G. (1550) *Lives of the Most Eminent Italian Architects, Painters and Sculptors*.

Weiler, M. & McDonnell, J. J. (2004) Virtual experiments: a new approach for improving process conceptualisation in hillslope hydrology. *J. Hydrol.* **285**, 3–18.

White, M. (2000) *Leonardo Da Vinci: the First Scientist*. Abacus, London, UK.

Whittow, J. B. (2000) *Dictionary of Physical Geography*. Penguin Books, London, UK.

Postscript

FROM LEONARDO'S LIVING EARTH CONCEPT TO EARTH SYSTEM SCIENCES

Some 500 years ago, Leonardo Da Vinci, largely relying on Plato's ideas, established a correspondence between the blood circulation in human bodies and the water cycle on earth. Leonardo was well aware that he faced a serious problem with the motion of water through the earth. Throughout his writings he struggled to discover the forces and processes that control the water cycle.

Leonardo was interested in understanding the processes that are the driving forces within his "living earth" concept in order to find new ways to protect human life and infrastructures from meteorological and hydrological extremes. For studying those processes, he progressively developed a systematic working sequence, based on the elaboration of a hypothesis, followed by detailed observations of natural processes and/or of experiments that he had designed previously, allowing him finally to verify his hypothesis.

Today, this way of proceeding is a universally recognised fundamental in every scientific discipline. With significant progress having been initiated only since the middle of the 19th century, hydrological sciences, like many other earth sciences, are still very young. Leonardo knew about the value of observations and experiments in the verification process of a hypothesis, saying that *"**whenever discussing water, you first have to invoke experience, before referring to reason**"*. As in many other domains, such as painting or sculpting, he was well ahead of his time in his approach to studying the hydrological cycle. His achievements appear even more outstanding when considering the very limited technical means that Leonardo had at his disposal. Even the high-tech tools for investigating hydrological processes available nowadays, such as isotope tracers, ERT instruments, space-borne remote sensing platforms, as well as the ever-increasing potential of computational resources, still serve in the end to verify concepts that were previously established.

Since Leonardo, our planet has undergone changes, sometimes irreversible, due to anthropogenic activities. The earth's surface has been remodelled much more in the past 200 years, than it had been over more than 20 centuries before. The ever-growing speed of socio-economic and technological development over the past 25 years has caused even more changes to the global environment than the industrial development that was initiated some 250 years ago. There is now major concern about potentially irreversible impacts for life on our planet due to the mainly non-sustainable exploitation of natural resources. It is thus of paramount importance to better understand the global functioning of our planet.

Earth system sciences refer to the interactions between the atmosphere, the hydrosphere, the geosphere and the biosphere. This recognition is perfectly illustrated by the *Amsterdam Declaration on Global Change* of 2001, stating that *"the earth system behaves as a single, self-regulating system, comprised of physical, chemical, biological and human components. The interactions and feedbacks between the component parts are complex and exhibit multi-scale temporal and spatial variability"*. To a certain extent, this modern concept is close to Leonardo's view of the earth, as resembling a living body that has blood, flesh and bones. Even certain modern conceptualisations of the earth system, such as John Lovelock's (1979), still controversial, Gaia theory, consider our planet as a living entity.

What in Leonardo's works was a kind of analogy has now become an integrative factor in modern science. The earth systems approach challenges habits of reductionism, and the tendency of many scientists to put their disciplines into boxes and ignore the progress made in other disciplines. In line with Leonardo's *Homo universalis* background, it becomes more and more important not only to look at the constituent elements of our environment, but to see the connections between them and understand how the system works. Leonardo is thus a supreme example of interdisciplinarity.

In any case, even though he still lacked so many pieces of the puzzle, Leonardo Da Vinci was still tremendously ahead of his time and thus close to the concept of the earth system. As we have seen beforehand, he had an emotional relationship with nature, for when he passed by marketplaces where birds were sold, he would buy the birds and take them out of their cages with his own hands, just to let them fly into the air and restore their liberty. In one sense, the "living earth" offered him a favour in return, that Vasari (1550) described as: *"nature was pleased so to favour him that, wherever he turned his thought, brain and mind, he displayed such divine power in his works, that, in giving them their perfection, no one was ever his peer in readiness, vivacity, excellence, beauty and grace"*.

Laurent Pfister
Head of the research unit 'Geohydrosystems and Landuse Management'
Department Environment & Agro-biotechnologies, Centre de Recherche Public – Gabriel Lippmann,
Belvaux, Grand-Duchy of Luxembourg

Lucien Hoffmann
Head of the Department 'Environment & Agro-biotechnologies'
Centre de Recherche Public – Gabriel Lippmann
Belvaux, Grand-Duchy of Luxembourg

List of Figures

Chapter 1 frontispiece	***The Arno River valley (5 August 1473).*** Gabinetto disegni e stampe della Galleria degli Uffizi, Firenze (Inv. 436E). © Alinari Archives/CORBIS.	1

Chapter 2 frontispiece	***Studies of flowing water, detail (c. 1509–1511). Note Leonardo's typical back-to-front (mirror) writing.*** The Royal Collection © 2008 Her Majesty Queen Elizabeth II (RL 12660v).	11
Fig. 2.1	***Representation of natural phenomena, detail (c. 1511–1512).*** The Royal Collection © 2008 Her Majesty Queen Elizabeth II (RL 12388r).	13
Fig. 2.2	***Studies of flowing water, detail (c. 1510–1915).*** The Royal Collection © 2008 Her Majesty Queen Elizabeth II (RL 12579r).	16
Fig. 2.3	***Studies of flowing water, detail (c. 1509–1511).*** The Royal Collection © 2008 Her Majesty Queen Elizabeth II (RL 12660v).	20

Chapter 3 frontispiece	***Heavy storm over riders and trees, detail (c. 1514).*** The Royal Collection © 2008 Her Majesty Queen Elizabeth II (RL 12376r).	25
Fig. 3.1	***Sketch of a whirlwind. Codex Leicester.*** © Seth Joel/CORBIS.	31
Fig. 3.2	***(A) Sketch of a wind vane by Leonardo Da Vinci (c. 1493–1494). Codex Paris***, Manuscript H. © RMN (Institut de France)/René-Gabriel Ojéda, Paris, bibliothèque de l'Institut. (B) Diagram of the equipment. (C) Illustration of wind acting on the equipment; V is wind velocity.	32
Fig. 3.3	***(A) Sketch of an anemometer by Leonardo Da Vinci (c. 1500). Codex Atlanticus.*** Copyrights Biblioteca Ambrosiana Auth No. Int 59/08. (B) Diagram of the equipment. (C) Illustration of the wind acting on the equipment; V is wind velocity.	32
Fig. 3.4	***(A) Sketch of a hygrometer by Leonardo Da Vinci (c. 1480–1486). Codex Atlanticus.*** Copyrights Biblioteca Ambrosiana Auth No. Int 59/08. (B) Diagram of the equipment at low atmospheric humidity; a contains a hygroscopic substance and b contains wax. (C) Diagram of how the equipment reacts to increased atmospheric humidity; the hygroscopic substance takes up water.	33
Fig. 3.5	***(A) Sketch of another hygrometer by Leonardo Da Vinci (c. 1483–1486). Codex Atlanticus.*** Copyrights Biblioteca Ambrosiana Auth No. Int 59/08. (B) Diagram of the equipment at low atmospheric humidity; a contains a hygroscopic substance and b contains wax. (C) Diagram of how the equipment reacts to increased atmospheric humidity; the hygroscopic substance takes up water.	33

Chapter 4 frontispiece	***Study of rock formations (approx. 1510–1513).*** The Royal Collection © 2008 Her Majesty Queen Elizabeth II (RL 12394r).	35
Fig. 4.1	*Diagram based on an illustration for a note on "the relative height of the surface of the sea to that of the land".*	40
Fig. 4.2	*Illustration of Leonardo's concept for the formation of mountains emerging from the water sphere.*	41
Fig. 4.3	***Detail showing currents in a meandering river (c. 1508–1512). Codex Leicester***. © Seth Joel/CORBIS.	43
Fig. 4.4	***Study of riverbed erosion and sedimentation (c. 1508–1512). Codex Leicester***. *Detail from sheet discussing river control techniques.* © Seth Joel/CORBIS.	44
Chapter 5 frontispiece	***Deluge over an alpine valley (c. 1506).*** The Royal Collection © 2008 Her Majesty Queen Elizabeth II (RL 12409r).	51
Fig. 5.1	***Exploding mountain: "water rushing out of a burst vein" (c. 1517–1518)***. The Royal Collection © 2008 Her Majesty Queen Elizabeth II (RL 12380r).	57
Fig. 5.2	*Diagram relating to Leonardo's notes on the circulation of water between the oceans and the tops of mountains (left), based on the concept of a siphon (right).* **Detail from the Drawing of Nude Figures and Siphons. Codex Paris**, *Manuscript G.* © Alinari Archives/CORBIS.	59
Fig. 5.3	***Device for the measurement of the travelling speed of wind or water, Codex Arundel***. Copyright © The British Library Board. All Rights Reserved, Arundel 263, f.241.	67
Chapter 6 frontispiece	***Discussion of water currents.*** From Sheet Discussing Water Currents, Codex Leicester. © Seth Joel/CORBIS.	71
Fig. 6.1	***Drawing showing a severe storm over a city in a mountainous area***. The Royal Collection © 2008 Her Majesty Queen Elizabeth II.	73
Fig. 6.2	***The eddy made by the Mensola, when the Arno is low and the Mensola full. Detail from Sheet Discussing the Confluence of Rivers, Codex Leicester***. © Seth Joel/CORBIS.	77
Fig. 6.3	***Hydraulics of the River Loire at Amboise. Codex Arundel***. *Copyright* © *The British Library Board. All Rights Reserved. Arundel 263, ff.264v, 269.*	79
Fig. 6.4	***Hatch sluice gate for a navigable canal. Codex Atlanticus***. *Copyrights Biblioteca Ambrosiana Auth No. Int 59/08.*	80
Fig. 6.5	***Drop-down sluice gates. Codex Atlanticus***. *Copyrights Biblioteca Ambrosiana Auth No. Int 59/08.*	81
Fig. 6.6	***Sluice gate hatch. Codex Atlanticus***. *Copyrights Biblioteca Ambrosiana Auth No. Int 59/08.*	82
Fig. 6.7	***Canal reinforcement gabion structure. Codex Paris***, *Manuscript L.* © RMN (Institut de France)/Gérard Blot. Paris, bibliothèque de l'Institut.	83

BENCHMARK PAPERS IN HYDROLOGY

A Series that collects together, by theme, the papers that provided the scientific foundations for hydrology in the 20th century. Published across a wide spectrum of disciplines, the papers define the field and provide an overview of the development of ideas that have led to current concepts and understanding. The Series Editor is Jeff McDonnell.

GROUNDWATER
Selection, Introduction and Commentary by Mary P. Anderson

Mary Anderson's selection of papers to reprint, and the commentaries that she has prepared to accompany them, document the development of groundwater hydrology in the 20th century. The topic is tackled in seven sections: **Establishing Fundamentals** includes a translation of Darcy's experimental results that led to the relation we know as Darcy's law, as well as classic papers by Meinzer, Theis, and Hubbert, among others. **Determining Parameters** covers the development of pumping test theory and field practice as well as approaches to estimating aquifer parameters in the field. **Flow System Analysis** includes Freeze & Witherspoon's (1967) frequently cited paper and a less well-known one by Theis, and others. **Parameter Uncertainty** research is represented by two papers (Freeze, 1975, and Marsily et al., 1984) that reflect the concern, which has yet to be resolved, regarding quantification of uncertainty. **Interaction with Surface Water** traces how, from the mid-20th century on, recognition of groundwater interaction with the ocean, lakes and streams grew and influenced related disciplines. **Contaminant Processes** contains papers that were significant to early research on contaminant occurrence and transport, i.e. in the 1970s and 1980s. **Dispersion and Heterogeneity** links Slichter's (1905) seminal contribution that identified dispersion in the field and Skibitzke & Robinson's (1963) laboratory findings, with more recent attempts to represent these phenomena with models.

Volume 3 *in the* IAHS Benchmark Papers In Hydrology Series

ISBN 978-1-901502-74-9 (2008) A4 format, ~600 pp. Provisional price, £50.00

 Sponsored by SAHRA, University of Arizona, USA

Please send book orders and enquiries to:

Mrs Jill Gash
IAHS Press, Centre for Ecology and Hydrology
Wallingford, Oxfordshire OX10 8BB, UK

jilly@iahs.demon.co.uk
tel.: + 44 1491 692442
fax: + 44 1491 692448/692424

Book prices include postage worldwide. IAHS Members receive discounts on IAHS publications.

See *www.IAHS.info* for information about IAHS (the International Association of Hydrological Sciences), membership, publications, meetings and other activities.

BENCHMARK PAPERS IN HYDROLOGY

A Series from IAHS that collects together, by theme, the papers that provided the scientific foundations for hydrology in the 20th Century. Published across a wide spectrum of disciplines, these papers define the field and provide an overview of the development of ideas that have led to our current concepts and understanding. The Series Editor is Jeff McDonnell.

STREAMFLOW GENERATION PROCESSES

Selection, Introduction and Commentary by Keith J. Beven

The 30+ papers reproduced span the period from 1933 to 1984, commencing with Horton's early papers on infiltration and on maximum groundwater levels. With the aid of the Introduction and Commentary, they provide a stimulating insight to developments in this part of the field of hydrology.

ISBN 978-1-901502-53-4 (2006) A4 format, softback, 432 + x pp. £40.00

EVAPORATION

Selection, Introduction and Commentary by
John H. C. Gash & W. James Shuttleworth

The development of evaporation measurement techniques are documented first, commencing with the Wagon Wheel Gap catchment water balance (1921), through mass budget to water transfer methods, and use of scintillometry. Dalton's seminal essay *On Evaporation* (1802) starts the selection of papers on evaporation estimation, which then covers atmospheric controls on the evaporation process (the original Penman and Thornthwaite papers are reproduced), vegetation controls via transpiration and interception, and finally evaporation as a component of the global climate system. The Commentaries explain the context and significance of each paper.

ISBN 978-901502-98-5 (2007) A4 format, softback, 526 + x pp., £40.00

Please send book orders and enquiries to:

Mrs Jill Gash
IAHS Press, Centre for Ecology and Hydrology
Wallingford, Oxfordshire OX10 8BB, UK

jilly@iahs.demon.co.uk
tel.: + 44 1491 692442
fax: + 44 1491 692448/692424

Prices include packing and postage. 25% discount available to IAHS members purchasing for their personal use.

 The IAHS Benchmark Papers in Hydrology Series is sponsored by SAHRA, the Center for Sustainability of semi-Arid Hydrology and Riparian Areas, University of Arizona.

Climate and the Hydrological Cycle

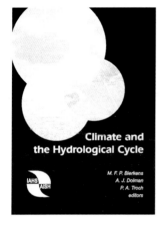

Edited by

Marc Bierkens Utrecht University, The Netherlands
Han Dolman Vrije Universiteit Amsterdam, The Netherlands
Peter Troch University of Arizona, USA

An in-depth overview of the role of the hydrological cycle within the climate system, including climate change impacts on hydrological reserves and fluxes, and the controls of terrestrial hydrology on regional and global climatology. This book, composed of self-contained chapters by specialists in hydrology and climate science, is intended to serve as a text for graduate and postgraduate courses in climate hydrology and hydroclimatology. It will also be of interest to scientists and engineers/practioners interested in the water cycle, weather prediction and climate change.

1. The Role of the Hydrological Cycle in the Climate System 2. Evaporation
3. Physics of Evaporation and Atmospheric Boundary Layers Over Land
4. Precipitation Physics and Rainfall Observation 5. Land Surface Hydrology
6. Land Surface Schemes and Climate Models 7. Arctic and Snow Hydrology
8. Dynamics of Glaciers, Ice Sheets and Global Sea Level
9. Feedback Mechanisms: Precipitation and Soil Moisture
10. Feedback Mechanisms: Land Use, Hydrology and Carbon
11. Palaeohydrology: An Introduction 12. Groundwater Palaeohydrology
13. Global Warming and the Acceleration of the Hydrological Cycle
14. Climate Change and Hydrological Impact Studies
15. Remote Sensing for Hydrological Studies

IAHS Special Publication 8

(*October 2008*) ISBN 978-1-901502-54-1 (Paperback); 344 + xvi pages
Price £50.00

Sponsored by:

Please send book orders and enquiries to:

Mrs Jill Gash
IAHS Press, Centre for Ecology and Hydrology
Wallingford, Oxfordshire OX10 8BB, UK

jilly@iahs.demon.co.uk
tel.: + 44 1491 692442
fax: + 44 1491 692448/692424

Book prices include postage worldwide. IAHS Members receive discounts on IAHS publications.

See *www.IAHS.info* for information about IAHS (the International Association of Hydrological Sciences), membership, publications, meetings and other activities.

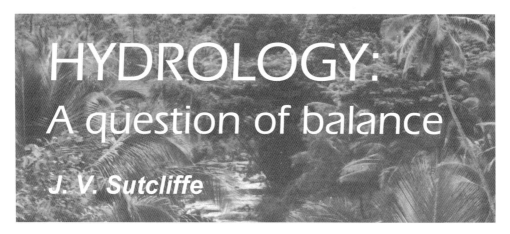

HYDROLOGY: A question of balance
J. V. Sutcliffe

IAHS Special Publication 7
(*November 2004*) ISBN 978-1-901502-77-0 (Paperback); 200 + xviii pages;
Price £30.00

The International Water Management Institute, Colombo, Sri Lanka, sponsored this publication

Hydrology: A Question of Balance is a unique hydrology text. It brings hydrological analysis to life by means of examples in which the author has been involved: numerous practical problems that had to be tackled (often despite limited data, resources and time) are described and the methods that were used to find a solution are explained. The application of a water balance is an essential component of solving these applied problems.

John Sutcliffe offers the experience of a hydrologist with extraordinary practical expertise, assembled in areas with different climates, topographies, levels of development, and cultures. Projects in many countries, including Sudan, Iran, Senegal, Botswana, India, Sri Lanka, New Zealand, Bosnia, Poland and the UK, are detailed to illustrate how hydrologists can, and need to, use all the available information to understand the hydrological context of their studies. Practising hydrologists and engineers, as well as students, will learn from this volume which complements standard hydrology textbooks.

Need for Hydrological Information • Network Design and Appraisal • Rainfall • Evaporation and Transpiration • Soil Moisture Storage • Groundwater Recharge • River Flow • Water Balance • Surface Water Resource Assessment • Flood Estimation • Sedimentation and Environmental Issues • Postscript

---

Please send book orders and enquiries to:

Mrs Jill Gash
IAHS Press, Centre for Ecology and Hydrology
Wallingford, Oxfordshire OX10 8BB, UK

jilly@iahs.demon.co.uk
tel.: + 44 1491 692442
fax: + 44 1491 692448/692424

See the IAHS website for information about the International Association of Hydrological Sciences (IAHS), membership, meetings, and publications, including abstracts of papers published since 2000.

Sediment Dynamics in Changing Environments

Edited by Jochen Schmidt, Tom Cochrane, Chris Phillips, Sandy Elliott, Tim Davies & Les Basher

IAHS Publ. 325 (*2008*) ISBN 978-1-901502-84-8, 626 + xiv pp. Price £105.00

Abstracts of the papers in this volume can be seen at: ***www.iahs.info***

with information about other IAHS publications and IAHS activities and membership

To understand *Sediment Dynamics in Changing Environments* we need to advance our knowledge of sedimentary processes and systems, and in particular of associated scaling issues. This knowledge, derived from information and analysis of historical sediment archives and system analysis and modelling, enhances our abilities to assess impacts of global change on sedimentary systems. Most importantly, we need to find ways to link our understanding and models of sedimentary systems with impacts on human environments, including hazard and risk assessment, improvement of management, and feedback into policy frameworks. The papers in this book, first presented at an IAHS Symposium in New Zealand in December 2008, document the international research efforts going into the themes of:

1. Unlocking the archives – dating and source tracing technologies
2. Processes and scales in sedimentary systems – from point to continents
3. Global change and erosion
4. Linking erosion with environmental and societal impacts

Please send book orders and enquiries to:

Mrs Jill Gash
IAHS Press, Centre for Ecology and Hydrology
Wallingford, Oxfordshire OX10 8BB, UK

jilly@iahs.demon.co.uk
tel.: + 44 1491 692442
fax: + 44 1491 692448

Groundwater Quality:
Securing Groundwater Quality in Urban and Industrial Environments

Edited by Michael G. Trefry
CSIRO Land and Water, Australia

IAHS Publ. 324 (*2008*) ISBN 978-1-901502-79-4, 566 + x pp. Price £90.00 (includes postage)

Selected and reviewed papers from GQ07, the groundwater quality conference held in Fremantle, Australia, in December 2007

Our relationship with groundwater is bipolar. Increasingly, we depend on it for our very survival, both in developed and developing nations. However, our urban and industrial activities involve routine and detrimental impacts to the quality of our groundwater reserves. The science of groundwater quality is therefore paramount to underpin successful and sustainable management of this precious resource. GQ07 focused on a range of urban and industrial groundwater quality issues, including:

- major instances of groundwater contamination and consequent human impact,
- emerging chemicals of concern and the ability of the environment to assimilate them,
- new contamination assessment, characterization and remediation techniques,
- data integration and analysis for decision making,
- development of water management policy and controls,
- groundwater quality transformations near receiving environments.

The research papers published here, by scientists and water professionals from around the globe, form a valuable summary of the state of knowledge in these areas. Key topics are arsenic contamination, management of radioactive sites, non-aqueous phase liquids, biogeochemical and isotopic processes, land-use influences, surface water–groundwater interaction, and the regulation and protection of groundwater supplies.

The abstracts of the papers can be seen at: **www.iahs.info** *with information about other IAHS publications and IAHS activities.*

Please send book orders and enquiries to:

Mrs Jill Gash
IAHS Press, Centre for Ecology and Hydrology
Wallingford, Oxfordshire OX10 8BB, UK

jilly@iahs.demon.co.uk
tel.: + 44 1491 692442
fax: + 44 1491 692448

River Basins – From Hydrological Science to Water Management

Bassins versants – de l'hydrologie à la gestion de l'eau

Edited by Ioulia Tchiguirinskaia, Siegfried Demuth & Pierre Hubert

IAHS Publ. 323 (*2008*) ISBN 978-1-901502-69-5, 154 + xii pp. Price £40.00 (includes postage)

A joint IAHS / UNESCO-IHP publication
proceedings of the Ninth Kovacs Colloquium, Paris, June 2008

The invited contributions provide a review of the practice and realities of undertaking research for river basin management (how to involve the public as stakeholders, building trust with decision makers, the research funding situation); the tools we have available (hydrological models, how good are they, how we can reduce uncertainties and explain them to policy makers); their application; and the current situation regarding water monitoring, research and management in El Salvador, India, Romania, Russia and South Africa. Conclusions and recommendations are summarized in a final section which proposes issues for future consideration in hydrological research and management.

De l'hydrologie du bassin à la gestion intégrée par bassin versant / From watershed hydrology to integrated watershed management   *Jean-Pierre Villeneuve et al.*

Bridging the gap between knowledge and policy action: land use is the key – confidence is the condition   *Giselher Kaule & Hans-Georg Schwarz-v.Raumer*

Overview of water resources systems modelling in South Africa   *Caryn Seago & Ronnie McKenzie*

Knowledge Management of water resources in El Salvador   *Ana Deisy Lopez Ramos*

Romanian national strategy for flood risk management   *Lucia Ana Varga et al.*

Sustainable water management by maintenance of the natural environment in river basins   *Elena Asabina*

Water-quality monitoring and process understanding in support of environmental policy and management   *Norman E. Peters*

Gestion intégrée et participative des ressources en eau: une perspective de sciences sociales / Integrated and participative river basin management: a social sciences perspective   *Bernard Barraqué*

The changing Indian scenario: from river basin studies to water management studies and its scientific rationale   *P. Rajendra Prasad*

Measurements, models, management and uncertainty: the future of hydrological science   *Keith Beven*

The abstracts of the papers can be seen at: **www.iahs.info** with information about other IAHS publications and IAHS activities.

Please send book orders and enquiries to:

Mrs Jill Gash
IAHS Press, Centre for Ecology and Hydrology
Wallingford, Oxfordshire OX10 8BB, UK

jilly@iahs.demon.co.uk
tel.: + 44 1491 692442
fax: + 44 1491 692448

 *www.iahs.info*

INTERNATIONAL ASSOCIATION OF HYDROLOGICAL SCIENCES

IAHS, established in 1922, is a nonprofit-making nongovernmental scientific organization.

IAHS is concerned with all components of the hydrological cycle and related processes, as well as the application of hydrology and its implications for society and the environment. Currently nine International Commissions deal with the hydrological cycle, water for society and the environment, and specific techniques:

- Surface Water
- Groundwater
- Continental Erosion
- Snow and Ice Hydrology
- Water Quality
- Water Resources Systems
- Coupled Land–Atmosphere Systems
- Remote Sensing
- Tracers

The Prediction in Ungauged Basins initiative (PUB, 2003–2012) involves all the Commissions, and several working groups address specific topics..

Information about IAHS, the Commissions, PUB, and their activities is available at www.iahs.info, or from the IAHS Secretary General:

Dr Pierre Hubert, Secretary General IAHS, UMR Sisyphe, Université Pierre & Marie Curie, Case 105, 4 Place Jussieu, 75252 Paris Cedex 05, France   e-mail: pjy.hubert@free.fr

Membership of IAHS is free of charge and open to anyone wishing to participate in IAHS activities, such as IAHS symposia and workshops. The worldwide membership is more than 5000.